Lectures on
STATISTICAL MECHANICS

by

M. G. BOWLER
Department of Nuclear Physics
Oxford University

PERGAMON PRESS

OXFORD · NEW YORK · TORONTO · SYDNEY · PARIS · FRANKFURT

U.K.	Pergamon Press Ltd., Headington Hill Hall, Oxford OX3 0BW, England
U.S.A.	Pergamon Press Inc., Maxwell House, Fairview Park, Elmsford, New York 10523, U.S.A.
CANADA	Pergamon Press Canada Ltd., Suite 104, 150 Consumers Road, Willowdale, Ontario M2J 1P9, Canada
AUSTRALIA	Pergamon Press (Aust.) Pty. Ltd., P.O. Box 544, Potts Point, N.S.W. 2011, Australia
FRANCE	Pergamon Press SARL, 24 rue des Ecoles, 75240 Paris, Cedex 05, France
FEDERAL REPUBLIC OF GERMANY	Pergamon Press GmbH, 6242 Kronberg-Taunus, Hammerweg 6, Federal Republic of Germany

First edition 1982.

Library of Congress Cataloging in Publication Data
Bowler, M. G.
Lectures on statistical mechanics.
Bibliography: p.
Includes index.
1. Statistical mechanics. I. Title.
QC174.8.B68 1982 530.1'3 81-15862
AACR2

British Library Cataloguing in Publication Data
Bowler, M. G.
Lectures on statistical mechanics.
1. Statistical mechanics
I. Title
530.1'3 QC174.8

ISBN 0-08-026516-2 (Hardcover)
ISBN 0-08-026515-4 (Flexicover)

In order to make this volume available as economically and as rapidly as possible the author's typescript has been reproduced in its original form. This method unfortunately has its typographical limitations but it is hoped that they in no way distract the reader.

Printed in Great Britain by A. Wheaton & Co. Ltd., Exeter

PREFACE

In a sense, all you need to know about equilibrium statistical mechanics is summarised in the magic formula

$$P \propto e^{-E/kT}$$

Every physicist knows this formula and uses it continually and usually unconsciously. Its origins and meaning are however too often shrouded in obscurity.

In 1978 I was asked to give a course on statistical mechanics, consisting of 16 lectures, to second year undergraduates reading physics in Oxford. Since I have never used statistical mechanics in my research (in experimental high energy physics) and had not studied the subject since I was myself an undergraduate, I had to go right back to the beginning and attempt to acquire an understanding. Though gruelling, it was tremendous fun and I hope that this volume of lectures will illuminate the foundations of this fascinating, but conceptually difficult, subject for other beginners.

In educating myself in Statistical Mechanics I was continually confused and irritated by most of the elementary treatments of the subject which seem too often obscure at crucial points even when not downright mendacious. I therefore determined to construct a course devoted primarily to the fundamental assumptions of equilibrium statistical mechanics and to a critical analysis of the methods leading from those fundamental assumptions to the magic formula, its variants, and use. My hope was to provide a solid foundation from which any of the multifarious developments and applications could be confidently explored.

Most students are shocked by the notion that statistics and probability have any fundamental role in physics, so in Lecture 1 I demonstrated that the classical concept of cross section is essentially statistical and then used it in a context where the importance of statistical assumptions is manifest. In Lecture 2 I applied the ideas of cross section and interaction rates to states of equilibrium, showing that the distribution of particles in a gas over the energy states available necessarily follows the magic formula, given only conservation of energy and the existence of an equilibrium. This elegant (and mathematically trivial) treatment does not seem to be widely known, which is a pity. The addition of the Pauli exclusion principle yields at once the Fermi-Dirac distribution function, and, with a little more quantum mechanics, the Bose-Einstein distribution also falls out. Having obtained the magic formula, the meaning of temperature is explored in the second lecture, with a digression on the subject of negative temperatures.

Lecture 3 introduced the traditional microcanonical treatment of statistical mechanics, but with careful attention to the fundamental assumptions and to the use of undetermined multipliers, while Lecture 4 was devoted to a critical analysis of this treatment which I hope clarifies the difficulties encountered. Lecture 5 was concerned with the distribution in energy of the components of a classical perfect gas and is very standard apart from a qualitative discussion of the validity of the density of states factor for arbitrarily shaped containers. Lecture 6 covered equipartition of energy and specific heats of gases and solids, including the freezing out of degrees of freedom. Lectures 7 and 8 established thermodynamics for a microcanonical ensemble and in particular identified the entropy function.

The fun started again in Lecture 9, dealing with the difficulties encountered in obtaining a consistent definition of entropy for a gas of identical particles and a critical discussion of how these difficulties may be resolved within a classical microcanonical treatment. But the solution really requires a discussion of identical particles in quantum mechanics and this was provided in Lecture 10 (for both unlocalised and localised particles). The Bose-Einstein and Fermi-Dirac distributions were extracted in the traditional way and the utter unreliability of the treatment emphasised.

Lecture 11 introduced the canonical and grand canonical treatments of equilibrium statistical mechanics. It seems to me essential that any introductory course in statistical mechanics should do this because of their great utility in specialised applications. But there are other reasons. The canonical treatment at least is necessary in order to establish thermodynamics from statistical mechanics for arbitrary macroscopic systems, and the grand canonical treatment allows clean extraction of the Fermi-Dirac and Bose-Einstein distributions. This is done in Lecture 13, following Lecture 12 which dealt with mean values and fluctuations and (I hope) cleared up some of the obscurities initially enveloping the canonical and grand canonical treatments.

Lecture 14 was devoted to bosons: the curious Bose-Einstein condensation and the treatment of black body radiation, including the Einstein A and B coefficients for spontaneous and stimulated emission of radiation. Lecture 15 applied the Fermi-Dirac distribution to the interiors of metals and stars.

Lecture 16 is divided into three parts, and is concerned with statistical mechanics in the context of gravitation. In Part I the trivial but illuminating problem of the isothermal atmosphere is treated. In Part II statistical mechanics is applied to star clusters, galaxies and clusters of galaxies. This required the introduction of the virial theorem and is only a quasi-equilibrium problem - star clusters evaporate. Part III is a brief introduction to the thermodynamics of black holes, a subject which suggests deep connections between quantum field theory, general relativity and that triumph of 19th century physics, thermodynamics.

PROBLEMS are scattered throughout the lectures and should be regarded as an integral part of the course. Some are very easy, some represent standard developments from preceding material, a few are very difficult. I hope they are all illuminating - and fun.

Typographical note: Where there is danger of confusing the lower case letter l with the numeral 1, I have written the former in script, as in ℓn for logarithm to the base e.

CONTENTS

Lecture 1

SOME STATISTICS AND A PROBLEM IN MECHANICS

Thermodynamics is an axiomatic theory of something called heat, defined either operationally as something you measure with thermometers, or by the laws of thermodynamics themselves. Statistical mechanics tells us what heat is, in terms of our knowledge of the structure of matter. It provides an explanation for the laws of thermodynamics in terms of the microscopic structure of matter, provides a recipe for the calculation of thermal properties which cannot themselves be calculated within thermodynamics, and in some cases the calculations can actually be carried through.

Statistical ideas pervade physics, even when they are not explicitly labelled as such. A trivial example is the notion of cross-section in classical physics. If we consider the relative motion of the Earth and a suitable asteroid - say Icarus - and we know the elements of each orbit, then we can predict with certainty whether or not Icarus will hit the Earth on a particular pass. The idea of a cross-section for this collision never enters the problem. But if we knew that the orbit of Icarus was subjected to certain perturbations which we did not know how to calculate, then we might have to say that on the next close approach Icarus could be found anywhere in a cross-section of a tube of radius 10^7 km. We then would not know whether or not a collision would take place but we would compute a probability for it - the cross-sectional area presented by the Earth divided by the cross-sectional area of the Icarus tube. ($P \sim 10^{-6}$ with the above number). This is a statistical idea, and cross-section is basically a concept of statistical physics. It isn't all that much use in the above example, for in the end the Earth is either hit (P = 1) or not hit (P = 0). We are attaching a number to our expectations: if we calculated $P \sim 0.1$ we would be worried about the future.

The idea is more directly useful if we consider the Earth moving through a cloud of interstellar gas and we want to know how much gas hits the atmosphere. We don't know the motion of the individual molecules in the cloud but we may perhaps know that the average density of the cloud is 10^3 molecules cm^{-3}, and that the relative velocity of the Earth and the cloud is 100 km sec^{-1}. With such a relative velocity we neglect the effect of the Earth's gravitational field (escape velocity 11 km sec^{-1}) and simply note that the Earth sweeps out a tube 100 km long and of cross-sectional area $\sim 10^8$ km^2 each second. The gas in a volume 10^{10} km^3 is swept up each second, which means $10^{10} \times 10^{15} \times 10^3 = 10^{28}$ molecules each second. (It doesn't correspond to much gas - if the cloud is mostly hydrogen only ~ 10 kg each second). This is much more useful - the error on 10^{28} molecules is 10^{14} molecules so we can compute (statistically) the number of molecules swept up to an accuracy of a few parts in 10^{14} (the result for mass collected each second becomes exact if the gas is replaced by an idealised continuous fluid).

Why is the error on our estimate of N molecules given by $\sqrt{N}$? In making our calculation of N we have supposed that each molecule has a certain probability $P\Delta V$ of being found in any volume element ΔV, subject to the constraint that the average density is ρ. (Let the cloud contain N molecules and be of volume V.) Then the probability that a volume ΔV contains molecules 1, 2, 3 ...n (that is, n labelled molecules) is simply

$$(P\Delta V)^n$$

However, we don't mind <u>which</u> n molecules are in ΔV: what is the probability that <u>any</u> n molecules are to be found in ΔV (out of a total of N)? We multiply the probability of a specified n molecules being in ΔV by the number of ways of choosing n molecules from among N. There are N ways of choosing the first, N-1 ways of choosing the second ... N - n + 1 ways of choosing the nth. The total number of ways of choosing our n molecules is

$$N(N - 1)(N - 2)\ldots(N - n + 1) = \frac{N!}{(N - n)!}$$

We are after the probability of finding a number of particles n in ΔV. So far we have a number

$$(P\Delta V)^n \frac{N!}{(N-n)!}$$

This is the probability of having n specified particles in ΔV, multiplied by the number of ways of choosing n particles from among N. But it isn't what we want. Suppose particles 1 and 2 are in ΔV. In our number of ways of choosing particles we first counted picking particle 1 first, then particle 2 AND we counted picking particle 2 first, then particle 1. All we care about is that particle 1 is there and so is particle 2. The number

$$\frac{N!}{(N-n)!}$$

is far too big. We must divide by the number of ways of ordering a specified n particles. An easy way of seeing the problem: if n = N then the number we get from $\frac{N!}{(N-n)!}$ is n!, whereas we want the number 1 because all the particles are now specified. The number of ways of ordering a specified n particles is easy: there are n ways of choosing the first, n - 1 ways of choosing the second ... so the answer is n!

Our number has now become

$$(P\Delta V)^n \frac{N!}{(N-n)!n!}$$

and this STILL isn't the probability we are after. It gives the probability of finding n particles in ΔV - but where are all the others? We want a number n in ΔV so we must specify that the rest of the particles are elsewhere - in the volume $V-\Delta V$. We must multiply our number by the probability that the remaining particles are in $V-\Delta V$. We have already done all the counting and permuting necessary, so our final answer for the probability of finding n particles only in ΔV, calculated on the assumption that each particle has a probability P per unit volume, independent of the location of that volume, is

$$P(n) = (P\Delta V)^n \{P(V-\Delta V)\}^{N-n} \frac{N!}{(N-n)!n!}$$

To make sure we haven't left anything out (it is easy to do) we make a check. The probability that either zero, or 1, or 2, ... or n, ... or N particles are in ΔV exhausts all possibilities and so should be

1. That is, if we have done our sums correctly

$$\sum_0^N P(n) = 1$$

Try it: $$\sum_0^N (P\Delta V)^n \{P(V-\Delta V)\}^{N-n} \frac{N!}{(N-n)!n!} = ?$$

It looks very nasty, but fortunately we know the BINOMIAL THEOREM

$$(x + y)^N = \sum_0^N x^n y^{N-n} \frac{N!}{(N-n)!n!}$$

This is easy to check for N a small number (like 1 or 2).

PROBLEM: Prove it in general - I did it by induction when preparing this lecture.

Then applying the binomial theorem our sum becomes

$$\{P\Delta V + P(V-\Delta V)\}^N = (PV)^N$$

But P (the probability per unit volume) is $^1/V$ so indeed the answer is 1.

So we have the probability of finding n particles in a volume ΔV. Find the maximum probability and then find out how much we have to change the corresponding number of molecules for the probability to drop by e(=2.718...).

How do we find the maximum of the beastly looking expression

$$P(n) = (P\Delta V)^n \{P(V-\Delta V)\}^{N-n} \frac{N!}{(N-n)!n!} \qquad ?$$

First, n is a large number so we treat it as a continuous variable. Secondly, it is convenient to maximise the logarithm - taking the logarithm breaks up the product into a sum.

PROBLEM: Prove that maximising $\ell n(f(x))$ with respect to x gives the same condition as maximising f(x).

$$\ell n P(n) = n\ell n(P\Delta V) + (N-n)\ell n\{P(V-\Delta V)\} + \ell n N! - \ell n(N-n)! - \ell n\, n!$$

What is ℓn N!? We want an expression we can easily differentiate.

$N! = N(N-1)\ldots 1$ and so an upper bound is clearly

$N! < N^N$ because there are N factors $\leqslant N$

This is almost good enough but not quite.

$$\ell n\, N! = \ell n N + \ell n(N-1) + \ell n(N-2) \ldots = N\ell n N - \frac{1}{N} - \frac{2}{N} \ldots$$

The correction term is clearly proportional to N (with other terms even smaller) and we need to get it right because we want to differentiate $\ell n\, N!$ and $\frac{\partial}{\partial N} N\ell nN = \ell nN + 1$, the last term being the differential of N.

We may resort to a very useful trick for summing a series which is a function of integers:

$$\Sigma F(n) = \Sigma F(n)\Delta n \qquad \text{since } \Delta n = 1$$

so
$$\Sigma F(n) \sim \int F(n)dn \qquad \text{for large n}$$

$$\ell nN! = \sum_{1}^{N} \ell nn \simeq \int_{1}^{N} \ell n\, n\, dn = [n\, \ell n\, n - n]_{1}^{N} = N\, \ell n\, N - N + 1$$

We don't care about the 1. It doesn't stay in differentiation or enter differences and is negligible in comparison with $N \sim 10^{28}$.

We shall use the result

$$\ell n\, N! \simeq N\, \ell n\, N - N \qquad \text{Stirling's slacker approximation*}$$

Use this result:

$$\ell nP(n) = n\, \ell n(P\Delta V) + (N-n)\ell n\{P(V-\Delta V)\} + N\ell n\, N-N$$
$$- (N-n)\ell n(N-n) + (N-n) - n\, \ell n\, n + n$$

For $n << N$ expand $\ell n(N-n) = \ell nN - {}^{n}/N \ldots$

so that

$$\ell nP(n) \simeq n\, \ell n(P\Delta V) + (N - n)\ell n\, P(V - \Delta V) + n\ell nN + n - n\ell nn$$

Differentiate and equate to zero:

$$\frac{\partial \ell nP(n)}{\partial n} = \ell n(P\Delta V) - \ell n\, P(V-\Delta V) + \ell nN - \ell nn = 0$$

So
$$\ell n\frac{N\Delta V}{n(V-\Delta V)} = 0 \qquad n \simeq N\frac{\Delta V}{V}$$

The result is hardly surprising!

Denote this equilibrium value by n_E.

Now consider $P(n_e) = \dfrac{P(n_E)}{e}$

or
$$\ell nP(n_e) = \ell nP(n_E) - 1$$

so
$$1 = (n_E - n_e)\ell n(P\Delta V) - (n_E - n_e)\ell n\, P(V-\Delta V)$$
$$+(n_E - n_e)\, \ell nN + (n_E - n_e) - n_E\ell nn_E + n_e\ell nn_e$$
$$= (n_E - n_e)\{\ell n(P\Delta V) - \ell n\, P(V-\Delta V) + \ell nN - \ell nn_E\}$$
$$+ (n_E - n_e) + n_e\ell nn_e - n_e\ell nn_E$$

---

* See Whittaker and Watson, Modern Analysis, Sec. 12.33 (4th Ed.).

The term in curly brackets is zero from the definition of n_E.

So $\quad 1 = (n_E - n_e) + n_e \ln n_e - n_e \ln n_E$

write $n_E - n_e = \Delta$

$$1 = \Delta + (n_E - \Delta)\ln(n_E - \Delta) - (n_E - \Delta)\ln n_E$$

Expand $\quad \ln(n_E - \Delta) = \ln n_E + \ln(1 - {}^{\Delta}/n_E)$

$$\ln(1-x) = -x - x^2/2 \ldots$$

so $\quad 1 = {}^{\Delta^2}/2n_E$

$\Delta = n_E - n_e = \sqrt{2n_E} \ldots$ The precise number in front of $\sqrt{n_E}$ depends on the precise definition of error.

So now we have justified saying that if we sweep up 10^{28} molecules each second on average, this number is accurate to ~ 1 part in 10^{14}. BUT it depends on the input. We assumed P constant, but there will be a gravitational field in the cloud. Or suppose the material in a cloud of known mean density was clumped together in grains. Then the mean number of molecules accumulated each second would still be $\sim 10^{28}$, but the error would have to be calculated on the mean number of grains. If each grain contained $\sim 10^{28}$ molecules then the error would be 100%. The existence of such strong <u>correlations</u> among the molecules would, if we didn't know about them, greatly disturb some aspects at least of the calculation.

PROBLEM: I have on my desk in front of me a set of 24 picture cards. They are so drawn that any side by side placing results in a harmonious landscape. The blurb on the packet says that there are

1686553615927922354187720

separate pictures that can be formed (you don't have to use all cards) and goes on to say that if everyone on the Earth thought of 1 combination each second it would take more than 16 million years to think of them all - check these assertions!

Lecture 2

STATES OF EQUILIBRIUM

In the first lecture we applied some statistical methods to a simple problem - the Earth moving through a cloud of interstellar gas. Our one assumption was that any molecule had the same probability P of being in any unit volume in the cloud and we never mentioned temperature at all because it wasn't that kind of problem. The interstellar cloud had a mean density of 10^3 molecules cm^{-3} and its temperature might be $\sim 20°K$. It might be internally in equilibrium, but not in equilibrium with the space around it which has a temperature $\sim 3°K$.

What do we mean by saying that such a cloud is in equilibrium? On a microscopic scale everything is changing all the time. Even at 20°K a hydrogen atom has a velocity ~ 1 km sec^{-1}. [PROBLEM: work it out from the results obtained in this lecture.] What we mean is that the density and velocity distribution of the gas molecules is unchanging with time, when averaged over a large number of molecules so that statistical fluctuations can be ignored. In this lecture we shall consider the distribution functions obtaining in such circumstances.

Suppose I have a <u>system</u> consisting of a number of <u>components</u>. I know the properties of the individual components (they might be molecules in a gas, spins in a magnetic field, stars in a cluster..) and my system has the components interacting weakly so that the individual properties are not significantly disturbed. In particular I know the energy levels of the components. It is both convenient and correct to assign to the components discrete energy levels, for energy is quantised and even for free particles we can pass to the continuum from a set of discrete states.

Let my system be homogeneous - each component has the same energy levels. These levels may be due to the internal structure of the components (like excited states in an atom) or due to the system itself (like kinetic energy of molecules confined in a box or the energy of spins in an imposed magnetic field) or both.

Now define a state of equilibrium as one in which the occupation numbers of the component energy levels are constant with time: we can clearly only expect this to be exact in the limit where a finite number of components is merged into an ideal fluid - so you can see where statistics is eventually going to come in again.

A state of the system would be completely specified (microscopically) by saying that component 1 is in component state x, component 2 is in component state y ...etc. If there are n_i components all in component state i, then n_i is the occupation number of component state i. Don't confuse the component energy levels with energy levels of the whole system.

How - microscopically - do we change an occupation number? If we have an isolated system (and we are for the moment considering an isolated system), only by interaction among the components. Consider as an example a gas of homogeneous composition. Then if N_i is the density of molecules in state i (proportional to the occupation number n_i for a fixed volume) a transition out of state i into state k is accompanied by a transition out of state j into state l and the rate is (per unit volume)

$$R_{ij \to kl} = v_{ij} N_i N_j \sigma_{ij \to kl} \tag{1}$$

where v_{ij} is the relative velocity of molecules in states i and j and $\sigma_{ij \to kl}$ is the cross-section. If you like you may simply write

$$v_{ij} \sigma_{ij \to kl} = T_{ij \to kl}$$

where $T_{ij \to kl}$ is a normalised transition rate: this provides the operational definition of a cross-section.

The total transition rate OUT of state i is given by

$$R_{i\ out} = N_i \sum_{jkl} N_j T_{ij \to kl} \tag{2}$$

and the total transition rate INTO state i is given by

$$R_{i\ in} = \sum_{jkl} N_k N_l T_{kl \to ij} \tag{3}$$

provided the transitions are permitted by the conservation laws. Equilibrium requires that

$$R_{i\ out} = R_{i\ in}$$

If $T_{ij \to kl} = T_{kl \to ij}$, which is essentially the assumption of time reversal invariance, believed to hold for all except the milliweak interactions, then our equilibrium condition can be written

$$N_i \Sigma_{jkl} N_j T_{ij \to kl} = \Sigma_{jkl} N_k N_l T_{ij \to kl} \quad \text{for ALL } i \tag{4}$$

This condition is enormously restrictive. Consider for a moment two pairs of component states (i,j), (k,l) in equilibrium. For this case (4) becomes

$$N_i N_j = N_k N_l \tag{5}$$

or

$$\ell n N_i + \ell n N_j = \ell n N_k + \ell n N_l \tag{6}$$

subject to the restriction of conservation of energy in the process

$$E_i + E_j = E_k + E_l \tag{7}$$

The N_i etc. are functions of the energies E_i and so (6) and (7) together require

$$N_i = \alpha e^{\beta E_i} \tag{8}$$

for indices i, j, k, l where α and β are constants independent of which of the four states we look at. We can now see that equation (4), together with conservation of energy, is solved by

$$N_i = \alpha e^{\beta E i}$$

once more and since i is a specified state (any specified state) the constants α and β are common to all states (because N_i is common to all terms on the left hand side of (4).) The only proviso is that our states are not degenerate: we could handle this complication with an appropriate weight factor.

So our equilibrium configuration for a system of weakly interacting components is given by the occupation number

$$n_i = \alpha e^{\beta E_i}$$

where n_i is the number of components in component energy level i with E_i. (Although I invited you to think of a gas, you can see that the same considerations apply to any system of weakly interacting components, for example spins in a magnetic field.)

There are, for an isolated system, additional constraints. The total number of components is N so

$$\sum_i \alpha e^{\beta E_i} = N \tag{9}$$

and the total energy of the system is E so

$$\sum_i \alpha E_i e^{\beta E_i} = E \tag{10}$$

and the constants α and β are determined by N and E - or vice versa. (In practice it is usual to give N and β)

Solving (9)

$$\alpha = \frac{N}{\sum_i e^{\beta E_i}}$$

$$n_i = \frac{Ne^{\beta E_i}}{\sum_i e^{\beta E_i}} = \frac{Ne^{\beta E_i}}{z} \tag{11}$$

where $z = \sum_i e^{\beta E_i}$ is an enormously important quantity, the single component partition function.

Let us now investigate the significance of β. Suppose for simplicity the energy levels E_i are uniformly spaced and very close, so that we can approximately write

$$\int \alpha e^{\beta E_i} dE_i = N$$

$$\int \alpha E_i e^{\beta E_i} dE_i = E$$

If the E_i have no upper bound (for example, particles in a box) then clearly β must be negative for these two integrals (or the corresponding sums) to be finite. Then

$$\alpha = -\beta N \text{ and } \frac{\alpha}{\beta^2} = E \text{ so } \beta = -\frac{N}{E} = -\frac{1}{\bar{E}_i}$$

we might venture to call the quantity $-^1/\beta$ a temperature.

Now note that if E_i has an upper bound, the sums are finite for positive β as well as negative β. The energy states of an atom of spin J in a magnetic field are 2J + 1 in number, and equally spaced, and so have such an upper bound. In such a system positive β is just as good as negative β: if β is positive the higher energy states are more populous and equilibrium will be maintained provided we only have spin-spin interactions and don't let them radiate away energy electromagnetically, for example. Such a situation is unstable, lasting only for a short time in comparison with the electromagnetic decay time of a spin in a magnetic field (or more realistically the spin-lattice relaxation time) but while it lasts it is a perfectly good equilibrium state, and no equilibrium lasts for ever. We can have negative temperatures! Systems with negative temperature have a lot of energy to get rid of, so they are very hot.

The application of our results to the theory of paramagnetism should now be obvious. Before we consider the properties of β in further detail, let us clear up some loose ends.

In deriving the result (8), (11) we assumed that only two body interactions mattered. Suppose, say, three body interactions were significant? You can see that by an extension of the arguments leading to (5)-(7) we would find

$$N_i N_j N_k = N_l N_m N_n \text{ with } E_i + E_j + E_k = E_l + E_m + E_n$$

and the same distribution function would come through.

There is another problem of much greater significance (which is never even mentioned in most elementary texts). Not only is energy conserved in an interaction between two components, so are the three components of momentum and so is angular momentum. Maybe we should write

$$N_i = \alpha e^{\beta E_i} e^{\sum_a \gamma_a K^a_i} \qquad (K^a_i = P_x, P_y, P_z, w_x, w_y, w_z) \qquad (12)$$

Now $K = \Sigma K_i N_i$ and and if $K^a = 0$ then γ_a must be zero because the K_i^a have both positive and negative values. If we work in a frame in which momentum and angular momentum are zero we may set all $\gamma_a = 0$ in (12) and thus return to (8) although in choosing such a frame we may find extra contributions to the energy: for example in a rotating space station there will be a gradient of atmospheric pressure increasing outwards to the rim.

Now we may return to our study of β. Suppose we take TWO systems, with N^a and N^b components, energies E^a and E^b. Allow them to interact with each other and exchange energy and let them come to thermal equilibrium (assuming they do come to thermal equilibrium).

Initially
$$N^a = \Sigma\alpha_a e^{\beta_a E_i^a}$$
$$E^a = \Sigma\alpha_a E_i^a e^{\beta_a E_i^a}$$

and similarly for system b.

Finally
$$N^{a+b} = \Sigma\alpha e^{\beta E_i} = N_a + N_b$$
$$E^{a+b} = \Sigma\alpha E_i e^{\beta E_i} = E_a + E_b$$

For simplicity, make our assumption of uniformly spaced energy levels when

$$\beta = -\frac{N^{a+b}}{E^{a+b}} = -\frac{N^a + N^b}{E^a + E^b} \qquad \beta_a = -\frac{N^a}{E^b} \qquad \beta_b = -\frac{N^b}{E^b}$$

So
$$\frac{N^a}{\beta_a} + \frac{N^b}{\beta_b} = -(E^a + E^b) = \frac{N^a + N^b}{\beta}$$

This property of $^1/\beta$ is just the property TEMPERATURE has and so we can indeed identify $-^1/\beta$ with temperature, on some suitable scale. PROBLEM: Show that energy has flowed from the hotter to the colder.

Now we can do something else cute. We assumed that the transition rate to state j did not depend on the occupation number n_j of that state. We know this is wrong: for identical particles of $\frac{1}{2}$ integral spin (e.g. electrons) there is the Pauli exclusion principle.

Let us suppose that the rate from i, j to k, l is given by

$$n_i n_j T_{ij \to kl}(1 \pm n_k)(1 \pm n_l)$$

Then our equilibrium condition is obtained from

$$n_i n_j (1 \pm n_k)(1 \pm n_l) = (1 \pm n_i)(1 \pm n_j) n_k n_l$$

whence
$$\frac{n_i}{1 \pm n_i}\,\frac{n_j}{1 \pm n_j} = \frac{n_k}{1 \pm n_k}\,\frac{n_l}{1 \pm n_l} \quad \text{with } E_i + E_j = E_k + E_l$$

and at once
$$\frac{n_i}{1 \pm n_i} = \alpha e^{\beta E_i}$$

Solve and obtain
$$n_i = \frac{\alpha e^{\beta E_i}}{1 \mp \alpha e^{\beta E_i}} \tag{13}$$

If we choose a + sign in (13) n_i is always less than 1, as required by the Pauli principle (NB: n_i is an average of course). This gives the Fermi-Dirac distribution function which we may write

$$n_i(\text{F-D}) = \frac{1}{e^{\beta(E_F - E)} + 1}$$

What about the other case, choosing the negative sign in (12)? If transition rates into a state are proportional to 1 + the occupation number, then we get

$$n_i(\text{B-E}) = \frac{1}{e^{a - \beta E} - 1}$$

which is the Bose-Einstein distribution function and applies to identical particles of integral spin.

[Reference: The Feynman Lectures on Physics, Vol. III, Ch. 4. Highly recommended. This explains why the transition rates into a state depend on the occupation number.]

Finally, it should be obvious that if any one of these systems is brought into thermal contact with a heat bath with a temperature given by β_H, then after equilibrium has been achieved it will have the appropriate distribution function with a β equal to β_H.

SUMMARY

We have now obtained essentially all the important results of statistical mechanics. The essential result is that the probability of a system having energy E_i when in thermal equilibrium is given by

$$P(E_i) = \frac{e^{\beta E_i}}{\sum_i e^{\beta E_i}}$$

where $z = \sum_i e^{\beta E_i}$ is the partition function. The quantity β is related to the temperature: usually β must be negative, $^1/\beta$ has the additive properties of a temperature and we can identify $^1/\beta = -kT$, where k is Boltzmann's constant (which depends on how we define our scale of energy and our scale of temperature).

Lecture 3

TRADITION

The distribution functions we obtained in the last lecture by considering equilibrium can also be obtained quite differently by a method that has further possibilities - the study of fluctuations. The MOST PROBABLE occupation numbers have the same distribution as the equilibrium occupation number, and this result hints at why complicated systems tend to equilibrium.

Consider as before a system made up of a large number of weakly interacting components. Each of the components has a set of energy levels E_i, even in interaction with the others. Let the system have a well defined number of components and a well defined energy. The latter condition means an energy lying between fairly well defined limits rather than a precise energy: we often consider systems in contact with heat baths, isolated systems are never isolated completely and finally there is the quantum mechanical uncertainty $\Delta E \Delta t \sim \hbar$.

A microstate of our system is defined by giving the energy level of each component - and with perhaps 10^{24} components this specification is impossible. The macroscopic properties of the system are determined by the number of components in each energy level, and the more ways there are of choosing microstates with the same number of components in each energy level, the more probable the (single) corresponding macrostate will be. In order to calculate the probabilities, we must make a statistical assumption. In the absence of any information to the contrary, we suppose that all microstates corresponding to a given total energy (or lying within the narrow limits on the energy) are equally probable. This is the fundamental assumption of statistical mechanics, and it clearly isn't true: conservation laws other than conservation of energy clearly forbid some microstates - for example an object isolated in space can't suddenly start moving off in one direction because all the molecular vibrations have lined up. But the number of allowed microstates is so enormous and the properties of the overwhelming majority are so similar that we can use this assumption and be confident that we are not introducing a serious error.

We now calculate the probability of a given macrostate corresponding to occupation numbers n_i, subject to the constraints

$$\sum_i n_i = N, \quad \sum_i n_i E_i = E \tag{1}$$

How many ways are there of picking n_i particles out of a total of N? We worked this out in the first lecture. We want n_i particles to be in the level E_i and the rest are to be somewhere else. Pick n particles out of N and there are

$$\frac{N!}{(N - n)!} \tag{2}$$

ways of doing it. Remember however that this counts picking out particle i before particle j AND picking out particle j before particle i, while physically we want i and j to be in the state we have chosen and this configuration to be counted once. There are n! ways of ordering n particles so the physically significant number of ways of picking n particles out of N is

$$\frac{N!}{(N - n)!n!}$$

In the first component energy level, let there be n_1 particles (or components) The number of ways of selecting them is

$$\frac{N!}{(N - n_1)!n_1!} \tag{3}$$

In the second level there are to be n_2 components. These must be selected from among $(N - n_1)$ components and so the number of ways of doing it is

$$\frac{(N - n_1)!}{(N - n_1 - n_2)!n_2!} \tag{4}$$

The number of ways of picking n_3 particles from among $N - n_1 - n_2$ is

$$\frac{(N - n_1 - n_2)!}{(N - n_1 - n_2 - n_3)!n_3!} \tag{5}$$

and so on ... We at last reach a level, say the kth, where the number of particles runs out when we pick n_k from among $(N - \sum_1^{k-1} n_i)$ and because this is the last level,

$$N - \sum_1^{k-1} n_i - n_k = 0$$

The number 0! is 1 (another funny property of that peculiar number 0) or if you like the number of choices is 1 when all other possibilities are exhausted. Our last term is therefore

$$\frac{(N - \sum_1^{k-1} n_i)!}{(N - \sum_1^{k-1} n_i - n_k)!n_k!} = \frac{(N - \sum_1^{k-1} n_i)!}{n_k!} \tag{6}$$

Now multiply all the terms together and define,

$$W = \frac{N!}{(N - n_1)!n_1!} \frac{(N - n_1)!}{(N - n_1 - n_2)!n_2!} \cdots = \frac{N!}{n_1!n_2!\ldots n_k!} \tag{7}$$

This finally gives us the total number of ways of arranging n_1 particles in level 1, n_2 in level 2 and so on: it gives us the total number of ways of making a macrostate specified by $n_1 \ldots n_k$.
[NB There may be other macrostates with the same total energy and we haven't counted them yet...] The number of ways of making our macrostate depends on each of the numbers $n_i \ldots n_k$ and has a very sharp maximum for a particular distribution of occupation numbers as a function of energy ... So we try to find the maximum of the function W, as a function of all the n_i's, subject to our two imposed constraints (1).

First take the logarithm

$$\ln W = \ln N! - \sum_i \ln n_i! \tag{8}$$

Apply Stirling's theorem for N, n_i large

$$\ln W = N\ln N - N - \sum_i (n_i \ln n_i - n_i) \tag{9}$$

[N.B. n_i may not be large - wait for the next lecture.]

The condition for a turning point in a function of i independent variables $F(x_i)$ is

$$\Delta F = \sum \frac{\partial F}{\partial x_i} \Delta x_i = 0$$

for arbitrary, small Δx_i

or

$$\frac{\partial F}{\partial x_i} = 0 \tag{10}$$

But we have two equations of constraint so our n_i are NOT all independent. We must therefore write

$$\frac{dF}{dx_i} = 0$$

where dF is the change which results from changing x_i including the effect of any constraints: that is require

$$\frac{\partial F}{\partial x_i} + \sum_{j\neq i} \frac{\partial F}{\partial x_j} \frac{\partial x_j}{\partial x_i} = 0 \tag{11}$$

Suppose our two equations of constraint are

$$G(x_i) = 0, \qquad H(x_i) = 0 \tag{12}$$

These hold for all values of x_i. Then

$$\delta G = 0 = \sum_j \frac{\partial G}{\partial x_j} \delta x_j$$

So we can write

$$\sum_j \frac{\partial G}{\partial x_j} \frac{\partial x_j}{\partial x_i} = 0 \tag{13}$$

which relates the quantities $\partial x_j/\partial x_i$ (simple example : $G = x + y - c$) We can therefore write

$$\sum_j \left(\frac{\partial F}{\partial x_j} \frac{\partial x_j}{\partial x_i} + \lambda\frac{\partial G}{\partial x_j} \frac{\partial x_j}{\partial x_i} + \mu\frac{\partial H}{\partial x_j} \frac{\partial x_j}{\partial x_i}\right) = 0 \tag{14}$$

where λ, μ are arbitrary constants and j can take any value including i. This equation is devoid of useful content, but I can write it

$$\sum_j \left(\frac{\partial F}{\partial x_j} + \lambda\frac{\partial G}{\partial x_j} + \mu\frac{\partial H}{\partial x_j}\right) \frac{\partial x_j}{\partial x_i} = 0 \tag{14a}$$

and (14a) is clearly satisfied by the condition

$$\frac{\partial F}{\partial x_j} + \lambda\frac{\partial G}{\partial x_j} + \mu\frac{\partial H}{\partial x_j} = 0 \tag{15}$$

There are i such equations and G and H are known so we can solve for F in terms of λ and μ and get λ and μ from the constraints, so finding F for the turning point subject to constraints.

In our case,

$$F = (N\ell nN - N) - \sum_i (n_i \ell nn_i - n_i)$$

$$G = \Sigma n_i - N$$

$$H = \Sigma n_i E_i - E$$

$$\frac{\partial F}{\partial n_j} = -\ell nn_j, \frac{\partial G}{\partial n_j} = 1, \frac{\partial H}{\partial n_j} = E_j$$

so (15) gives us

$$-\ell nn_j + \lambda + \mu E_j = 0; \quad n_j = e^{\lambda} e^{\mu E_j}$$

and renaming the arbitrary constants

$$n_i = \alpha e^{\beta E_i}$$

once more, where the values of α and β which satisfy the whole system of equations may be obtained from

$$\Sigma \alpha e^{\beta E_i} = N, \ \Sigma \alpha E_i e^{\beta E_i} = E$$

The most probable distribution of occupation numbers is the same as the equilibrium distribution we obtained from quite different considerations.

The method we have just used to find the occupation numbers at the (constrained) maximum is called the method of undetermined multipliers (the multipliers λ and μ are only undetermined at the beginning of the game, not at the end). The recipe is simple: to find a turning point in F subject to the constraints G = 0, H = 0, ...

Instead of writing

$$\frac{\partial F}{\partial x_i} = 0, \text{ write}$$

$$\frac{\partial F}{\partial x_i} + \lambda \frac{\partial G}{\partial x_i} + \mu \frac{\partial H}{\partial x_i} .. = 0$$

Solve and then find λ and μ from the constraint equations. Note that the equations

$$\sum_j \left(\frac{\partial F}{\partial x_j} + \lambda \frac{\partial G}{\partial x_j} + \mu \frac{\partial H}{\partial x_j}\right) \frac{\partial x_j}{\partial x_i} = 0 \tag{14a}$$

are true at a turning value regardless of λ, μ. The values of $\partial x_j / \partial x_i$ are NOT arbitrary, so we equate their coefficients to zero as a convenience, not because (14a) forces us to. We have chosen a sufficient condition to satisfy (14a) which yields equations for F and there are just enough equations to determine F, λ and μ, so we have put in the constraints.

Suppose we were going to do the whole thing by brute force instead of being clever. Then we would write

$$\delta F = -\Sigma \ell n n_i \delta n_i$$

and add constraints $\Sigma \delta n_i = 0$, $\Sigma E_i \delta n_i = 0$ and solve for two of the n_i in terms of the others. It doesn't matter which two we take, so write

$$\delta n_1 + \delta n_2 = -(\delta n_3 \ldots \delta n_k)$$

$$E_1 \delta n_1 + E_2 \delta n_2 = -(E_3 \delta n_3 \ldots E_k \delta n_k) \tag{16}$$

Then

$$\delta n_2 (E_2 - E_1) = -(\delta n_3 (E_3 - E_1) \ldots \delta n_k (E_k - E_1))$$

$$\delta n_1 (E_1 - E_2) = -(\delta n_3 (E_3 - E_2) \ldots \delta n_k (E_k - E_2))$$

when

$$\delta F = -\ell n n_1 \{-\delta n_3 \frac{(E_3 - E_2)}{E_1 - E_2} \quad \ldots \quad \delta n_k \frac{(E_k - E_2)}{E_1 - E_2}\}$$

$$-\ell n n_2 \{-\delta n_3 \frac{(E_3 - E_1)}{E_2 - E_1} \quad \ldots \quad \delta n_k \frac{(E_k - E_1)}{E_2 - E_1}\}$$

$$-\ell n n_3 \delta n_3 \ \ldots \ \ell n n_k \delta n_k = 0$$

where $\delta n_3 \ldots \delta n_k$ are now arbitrary. Therefore

$$\ell n n_1 \frac{E_3 - E_2}{E_1 - E_2} + \ell n n_2 \frac{E_3 - E_1}{E_2 - E_1} - \ell n n_3 = 0 \tag{17}$$

(and for 1, 2, 3 we can substitute any indices we like).

$$\text{So } n_1^{E_2 - E_3} n_2^{E_3 - E_1} = n_3^{E_2 - E_1} \tag{18}$$

and we see at once that these equations are satisfied by

$$n_i = e^{\lambda + \mu E_i}$$ [PROBLEM: Check this!]

But the method is not as convenient as using undetermined multipliers.

Now a point which will be of importance later. We have assumed so far that the component energy levels are non-degenerate, and that we are not grouping levels in close bunches. Suppose that we have degenerate levels, or that we want to group them in close bunches about a mean E_i. What difference does that make? Take our present distribution

$$n(E_i) = \alpha e^{\beta E_i}$$

$$n(E_i + \Delta) = \alpha e^{\beta (E_i + \Delta)}$$

Then $n(E_i) + n(E_i+\Delta_1) + n(E_i+\Delta_2) \ldots = \alpha e^{\beta E_i}\{1 + e^{\beta\Delta_1} + e^{\beta\Delta_2} \ldots\}$

so that as $\Delta \to 0$

$$n(E_i) \to g_i \alpha e^{\beta E_i}$$

where $n(E_i)$ is now the total occupation of all levels with energy E_i and g_i is the degeneracy.

If you want to do it from the beginning, go back to the number of ways of choosing n_1 particles from N, which is

$$\frac{N!}{(N - n_1)!}$$

How many ways are there of distributing these n_1 particles over g_1 distinct levels corresponding to the same energy? This is trivial. There are g_1 ways of placing each particle so the total number of ways of placing n_1 particles is $g_1^{n_1}$ and we now must maximise

$$W = N! \Pi \frac{g_i^{n_i}}{n_i!}$$

So $$\frac{\partial \ell n W}{\partial n_i} = \ell n g_i - \ell n n_i$$

and on imposing the constraints

$$n_i = g_i \alpha e^{\beta E_i} \text{ once more.}$$

Note here that while n_i must be large, g_i need not be.

PROBLEM: Consider a gas at STP contained in a box $1m^3$. Calculate

(a) the probability all N molecules are in one half of the box;

(b) the probability a specified half of the molecules (numbers 1 to $^N/2$) are in one half of the box and the other specified half of the molecules are in the other half of the box.

(Calculate numbers, not just expressions.)

Lecture 4

CRITICAL ANALYSIS

In the third lecture we introduced the idea of microstates of our system of weakly interacting components, each microstate being specified by giving the state of each of the components. The macroscopic properties of the system are determined by the occupation numbers of the component energy levels and so a macrostate is specified by giving these occupation numbers. A huge number of microstates correspond to each macrostate. We calculated the number of ways of making up a macrostate characterised by a set of occupation numbers n_i

$$W = N!/\Pi\, n_i!$$

and maximised it subject to the constraints $\Sigma n_i = N, \Sigma n_i E_i = E$. This gave us the most probable distribution of occupation numbers and the result was identical to the (Boltzmann) equilibrium distribution. This suggests that we may take the most probable distribution calculated in this way as the equilibrium distribution. But before we can do this with a clear conscience, there are some difficulties to be cleared up. First, we calculated the maximum using Stirling's theorem, which is only valid for large n_i, and n_i is not necessarily large. Secondly, there are other macrostates, with different distributions, with the same total energy E (or lying indistinguishably close to E.) The most probable distribution is only representative if it is overwhelmingly the most probable or if all macrostates with comparable probability have essentially the same physical properties.

If all microstates of the same energy are equally probable, the most representative distribution of occupation numbers is given by the average over all macrostates

$$\hat{n}_i = \frac{\sum_j n_i w_j(n_i)}{\sum_j w_j(n_i)} \tag{1}$$

The most probable distribution function $\bar{n}_i$ is given by

$$\bar{n}_i = \frac{Ne^{-E_i/kT}}{\Sigma e^{-E_i/kT}} \tag{2}$$

(identifying the constant β with $-^1/kT$). If there are a relatively small FINITE number of energy levels, a lot of particles and T not too small, then ALL $\bar{n}_i$ will be large numbers and the use of Stirling's theorem is justified a posteriori. Suppose that lots of the n_i's are large, but there is a high energy tail where they are either zero or 1. In this case, since $0! = 1! = 1$ this tail won't much affect the value of a single $W(n_i)$ and we may expect Stirling's theorem to give us respectable results, applied to the number of ways of constructing the most probable macrostate,

$$W_{max}(\bar{n}_i)$$

For systems where Stirling's theorem may be applied, study the variation of the weight W as we go to other macrostates characterised by

$$n_i = \bar{n}_i + \delta n_i$$

We then have

$$\ell nW(n_i) = N\ell nN - N - \sum_i (n_i \ell n n_i - n_i)$$

$$= N\ell nN - \sum_i n_i \ell n n_i \tag{3}$$

Expand about $\bar{n}_i$:

$$\ell nW(n_i) = N\ell nN - \sum_i (\bar{n}_i + \delta n_i)\ell n(\bar{n}_i + \delta n_i)$$

$$= N\ell nN - \sum_i (\bar{n}_i + \delta n_i)\{\ell n\bar{n}_i + \ell n(1 + \frac{\delta n_i}{\bar{n}_i})\}$$

$$= N\ell nN - \sum_i (\bar{n}_i + \delta n_i)\{\ell n\bar{n}_i + \frac{\delta n_i}{\bar{n}_i} - \frac{1}{2}(\frac{\delta n_i}{\bar{n}_i})^2 \, ..\}$$

$$= \ell nW_{max}(\bar{n}_i) - \sum_i \{\underline{\delta n_i \ell n\bar{n}_i + \delta n_i} + \tfrac{1}{2}\frac{(\delta n_i)^2}{\bar{n}_i} \, ...\} \tag{4}$$

The sum of the two underlined terms is zero, because of the constraints $\Sigma n_i = N$, $\Sigma n_i E_i = E$. ($\Sigma \delta n_i = 0$ and $\Sigma \delta n_i E_i = 0$. Now $\bar{n}_i = \alpha e^{\beta E_i}$ so that $\ell n\bar{n}_i = \ell n\alpha + \beta E_i$ and $\Sigma \delta n_i \ell n\bar{n}_i = 0$). Then the weight of any macrostate other than the most probable is given by

$$W \simeq W_{max}(\bar{n}_i) \quad \exp\{-\tfrac{1}{2}\sum_i \frac{(\delta n_i)^2}{\bar{n}_i}\} \tag{5}$$

If each $\bar{n}_i$ is a large number, minute fractional excursions of n_i kill the relative probability of the corresponding macrostate and we may indeed take the most probable distribution as representative: the maximum is very sharp.

PROBLEM: You now have all the statistical mechanics you need to develop the theory of paramagnetic susceptibility. Do it!

The total number of ways of making macrostates consistent with the constraints is given approximately by

$$\Sigma W \sim W_{max}\int e^{-\delta n_1{}^2/\bar{n}_1} d\delta n_1 \int e^{-\delta n_2{}^2/\bar{n}_2} d\delta n_2 \ldots \int e^{-\delta n_i{}^2/\bar{n}_i} d\delta n_i \tag{6}$$

(Since the exponential of a sum is a product of exponentials and we approximately replace the sum over W by a multiple integral)

so

$$\Sigma W \sim W_{max}\prod_i (\bar{n}_i)^{\frac{1}{2}} \tag{7}$$

where we stop the product once $\bar{n}_i$ becomes ~ 1 if necessary. If we have G energy levels ($G \ll N$)

$$\Sigma W \sim W_{max}({}^{N}/G)^{G/2} \quad \text{and so}$$

$$\ell n(\Sigma W) \sim \ell n W_{max} + ({}^{G}/2)\ell n N - ({}^{G}/2)\ell n G \tag{8}$$

Terms of order $\ell n N$ were dropped in using Stirling's theorem, so note that to the accuracy of Stirling's theorem,

$$\ell n(\Sigma W) \sim \ell n W_{max} \quad \text{and this is good enough WHEN } G \ll N$$

This is no longer true if ${}^{N}/G$ is a small number and $G \sim N$.

So now go to a different situation where most values of $\bar{n}_i$ are ~ 1 or smaller. (Occupation numbers are integers so fractional values come from treating the n_i as continuous variables.) If most values of $\bar{n}_i$ are ~ 1 or smaller, this must mean that there are huge numbers of macrostates with $n_i = 0$ or 1, all pretty much indistinguishable physically as you swap 0's and 1's around. The sum of the total number of ways of making macrostates

$$\Sigma W$$

will have lots of terms $\sim N!$ and cannot be replaced by its maximum term.

Further, we cannot use Stirling's theorem anyway in finding even the maximum term.

What we can do is this. The situation in which most $n_i \lesssim 1$ will arise when the component energy levels are very closely spaced. Group levels together at approximately the same energy so that the total occupation of a group of g_j levels is m_j, m_j and g_j both large. Now calculate the most probable distribution of the m_j, using Stirling's theorem, and obtain

$$\bar{m}_j = \frac{N g_j e^{-E_j/kT}}{\Sigma g_j e^{-E_j/kT}} \tag{9}$$

The denominator is equal to $z = \sum_i e^{-E_i/kT}$ (counting every state)

and the value of $\bar{n}_i$ averaged over the group of levels in which level i is to be found is now

$$\bar{\bar{n}}_i = \frac{\bar{m}_i}{g_i} = Ne^{-E_i/kT}/z \tag{10}$$

which is the same answer we got for a system with well populated energy levels. It is quite clear that this treatment is a way of finding the average value of n_i over a large number of different (but similar) macrostates, and the term

$$W_{max}(g) = N! \Pi \left(\frac{g_j^{m_j}}{m_j!}\right)$$

satisfies $\ell n W_{max}(g) \simeq \ell n \Sigma W$ to terms of order $\ell n N$ (that is, to the accuracy of the form of Stirling's theorem we employ.) Thus the original treatment makes two compensating errors:

1. Using Stirling's theorem and
2. Taking the maximum term of the set of $W(n_i)$ as equal to the sum.

The important point for statistical mechanics is that all states with significant probability have essentially the same physical properties and distributions of occupation numbers which do not vary significantly from the Maxwell-Boltzmann distribution for large values of N.

For heavily populated component levels, $\delta n_i \sim \sqrt{\bar{n}_i}$ in any one level. For many levels with the same fractional deviation $|\delta n_i| = \bar{n}_i \delta$ we find

$$\Sigma(\delta n_i)^2/\bar{n}_i = \Sigma \delta^2 \bar{n}_i = \delta^2 N$$

and the relative probability of such a macrostate is down by a factor $e^{-\delta^2 N}$, requiring $|\delta n_i| \lesssim \bar{n}_i/\sqrt{N}$ or else the weight factor gets very small.

In systems where the levels are sparsely populated, occupation numbers 0, 1, 2... the individual fractional variations will be large ($\sim$1) but variations averaged over a group of closely spaced levels are small and the macroscopic behaviour is still given by the Maxwell-Boltzmann distribution.

We may look at the problem in a different way.
If I write

$$\ell nW = \ell nN!\,\Pi\left(\frac{g_i^{m_i}}{m_i!}\right)$$

$$\simeq N\ell nN + \Sigma m_i \ell ng_i - \Sigma m_i \ell nm_i$$

and set $m_i = Ng_i e^{-E_i/kT}/z$, then

$$\ell nW_{max}(g) = N\ell nN + \frac{N}{z}\Sigma\{g_i e^{-E_i/kT}\}\{\ell ng_i - \ell n(Ng_i e^{-E_i/kT}/z)\}$$

$$= N\ell nN + \frac{N}{z}\Sigma\{g_i e^{-E_i/kT}\}\{-\ell nN + E_i/kT + \ell nz\}$$

$$= E/kT + N\ell nz$$

and this result is independent of the value of the g_i's because of the compensating errors. Finally, with

$$\bar{n}_i = \bar{\bar{n}}_i = Ne^{-E_i/kT}/z$$

we identify $P_i = \bar{n}_i/N = e^{-E_i/kT}/z$ with the probability of finding any one particle (or component) in the level E_i (because multiplying by the total number of particles gives the number in the level). The probability of a particular microstate is thus

$$\Pi(P_i)^{n_i} \quad \text{and} \quad \sum_{\text{all microstates}} \Pi\,(P_i)^{n_i} = 1$$

and this must include all states not counted in W_{max}. If all the microstates that count have essentially the same probability P, their number is $^1/P$ and so we expect

$$\Sigma W \sim \frac{1}{\Pi(P_i)^{\bar{n}_i}} = \frac{1}{\Pi(P_i)^{NP_i}}$$

Then $\ln\Sigma W \sim -\Sigma NP_i \ln P_i = E/kT + N\ln z$ again! Thus P_i is the probability that any one particle is to be found in state E_i, taken over not just the most probable macrostate, but essentially all macrostates. Despite the defects of the treatment given in the last lecture, we may now happily accept the results - so far, so good!

[There is a much more satisfactory way of calculating n_i, or other averages, the method of Darwin and Fowler. It is mathematically much more complicated, requiring the multinomial theorem, Cauchy's theorem and the method of steepest descents. It may be found, for example, in the article by Guggenheim in Handbuch der Physik, Vol. III.]

A final point. Suppose we have two levels only (spin $\frac{1}{2}$ in a magnetic field) and set $E_1 = 0$, $E_2 \sim kT$. Then

$$z = 1 + e^{-1}$$

and with N particles $\ln\Sigma W \sim N$. The total number of microstates is thus

$$\Sigma W \sim e^N \text{ and } N \sim 10^{23} \text{ for a macroscopic specimen.}$$

A contribution to $\ln\Sigma W$ of order $\ln N$ is utterly negligible in comparison with N.

Suppose the system fluctuates from one microstate to another with spin flip frequency $\sim 10^{23}(=N) \times 10^{12}(={}^1/\tau) = 10^{35}\ \text{sec}^{-1}$. The time taken to explore all microstates of this simple system is thus $e^{10^{23}}/10^{35}$secs, which is forever. [See Kittel Thermal Physics, Ch. 4.] The point of the latter calculation is that even if we knew that a time average was equivalent to an average over microstates (the ergodic problem) we would still have to make a statistical assumption to get answers to a real physical problem, since in any experiment a macroscopic system explores only a negligible proportion of the states available.

Lecture 5

THE CLASSICAL PERFECT GAS

Having sorted out some obscurities in the application of the method of calculating the most probable distribution to systems with thinly populated component energy levels, we may now tackle the simplest example, the classical perfect gas. Such an ideal gas consists of pointlike particles, occupying no volume and having no internal structure and hence no internal excited states. These particles are confined within a box and unless we include the effects of external fields, such as gravitation, the energy of each particle is wholly determined by its kinetic energy and this energy is quantised: the energy levels of the components are the energy levels of a particle confined in a box. Finally we must allow weak interactions among these particles so that they can scatter and eventually an equilibrium will emerge, but the interaction must be weak enough not to significantly disturb the single particle energy levels. (There is no example of an ideal gas but some dilute real gases behave approximately as an ideal gas.)

Our first task is to enumerate the energy levels available to a particle confined in a box, for we can pass to the classical limit by replacing a discrete distribution by a continuous one. In determining the energy levels, the important thing is that the components of momentum are quantised because the spatial coordinates are restricted: the box is supposed to have impenetrable walls.

The boundary condition to be applied is that the wave function of the particle must vanish at the walls, for if it didn't the gradient of the wave function would be infinite at the walls, corresponding to infinite momentum. This boundary condition is analogous to the boundary condition appropriate for waves on a stretched, plucked, string - where the normal modes correspond to the length of the string being an integral number of half wavelengths (and the normal modes are the Fourier components of any more general disturbance permitted by the equation of motion and the boundary conditions). Suppose that a particle is confined in a cubic box. The Schrodinger equation is

$$\frac{p^2}{2m}\psi = E\psi \tag{1}$$

and for a cubic box we can separate ψ into functions of x, y, z and apply the boundary conditions to each function separately.

$$\psi \sim e^{\pm\frac{i}{\hbar}P_x x}\, e^{\pm\frac{i}{\hbar}P_y y}\, e^{\pm\frac{i}{\hbar}P_z z} \quad ; \quad E = (P_x{}^2 + P_y{}^2 + P_z{}^2)/2m$$

and $\frac{2\pi}{\lambda} = k = \frac{P}{\hbar}$

The solutions which vanish at the walls are clearly

$$\psi \sim \sin\frac{P_x x}{\hbar}\sin\frac{P_y y}{\hbar}\sin\frac{P_z z}{\hbar} \tag{2}$$

such that

$$\frac{P_i}{\hbar} L = n_i \pi \tag{3}$$

Define

$$n_x{}^2 + n_y{}^2 + n_z{}^2 = N^2 = \left(\frac{PL}{\hbar\pi}\right)^2 \tag{4}$$

so that

$$E = \frac{p^2}{2m} = \left(\frac{\hbar\pi}{L}\right)^2 \frac{N^2}{2m} \tag{5}$$

The number of states with energy E_N is equal to the number of ways of choosing (positive) integers n_x, n_y, n_z such that the sum of their squares is equal to N^2. The number of states with energy less than E_N is the number of ways of choosing three integers such that the sum of their squares is less than N^2.

[The one dimensional example is informative (the stretched string problem) when we have

$$kL = n\pi,$$

there is no degeneracy and the number of states with wave vector $\lesssim k$ is simply $n = {}^{kL}/\pi$.]

We can enumerate the modes in the three dimensional case by a bit of trivial coordinate geometry. Draw up a lattice of points with unit spacing. Every point (n_x, n_y, n_z) corresponds with one point on this lattice:

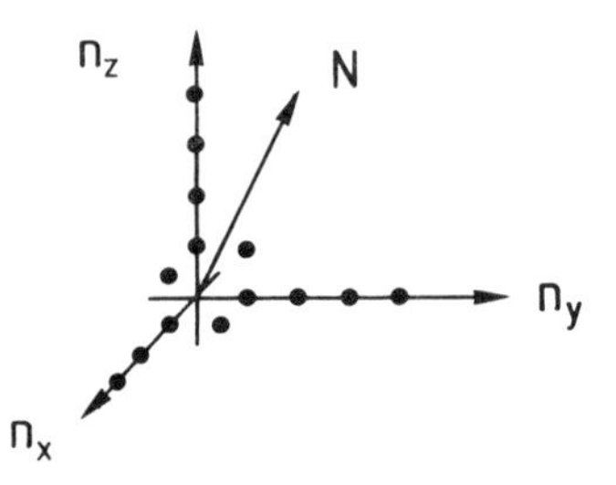

Draw a sphere of radius N. Then the number of lattice points which lie on its surface is the number of states with energy E_N (but we restrict ourselves to the positive octant because negative values of n_x, n_y, n_z do not give rise to different solutions, just a different arbitrary phase factor in front of the solution for positive n's). We can now find the number of states of energy $\leq E_N$ rather easily. The volume of a sphere of radius N is

$$\frac{4\pi}{3} N^3 = \frac{4\pi}{3} \left(\frac{P}{\pi\hbar}\right)^3 L^3$$

so the number n of states of energy less than or equal to E_N is given by the number of lattice points lying within the positive octant:

$$n = \frac{1}{8}\frac{4\pi}{3}N^3 = \frac{4\pi}{3}\left(\frac{P}{2\pi\hbar}\right)^3 L^3 = \frac{4\pi}{3}\left(\frac{P}{2\pi\hbar}\right)^3 V \tag{6}$$

if P is the magnitude of the momentum corresponding with E_N.

The number of states Δn lying between E_N and $E_N + \Delta E$ is given by the number of lattice points lying between two spheres of radii N and $N + \Delta N$ which is just

$$4\pi N^2 \Delta N \left(\times \frac{1}{8} \text{ for the positive octant}\right)$$

or

$$dn = 4\pi\frac{P^2 dP}{(2\pi\hbar)^3}V \tag{7}$$

where V is the volume of the box.

The results (6) and (7) are of enormous importance in both quantum mechanics and in statistical mechanics and must be understood.

Even though we obtained this result for a cubic box it is generally true that the density of states depends linearly on the volume through (6), (7) regardless of the shape of the confining box, provided only that the wavelength ($^h/P$) is very much smaller than any important dimension of the box. It can be demonstrated easily for a few high symmetry cases. [PROBLEM: Prove the result for a general rectangular parallelopiped. If you find that too easy, do it for a box whose cross-section is a right isosceles triangle.] The general proof is mathematically advanced and books on physics usually state the result and pass on: some are potentially misleading because they say that any volume can be approximately replaced by an assembly of little cubes. This could be taken to imply that the normal modes in an arbitrary volume are the normal modes of the little cubes which fill it, and this is quite false. A simple example will demonstrate the fallacy. Consider the modes on a stretched string. The number with a wave number $<k$ is $n(<k) = kL/\pi$. Now pin the string in the middle. The number of modes on each half (the halves now being independent) is $kL/2\pi$, which counted twice to give the total number of modes is kL/π once more. BUT all the odd modes on length L have been deleted by pinning the string in the middle and the even modes have been counted twice. The number is the same but the modes are not.

The proof that the density of states formulae (6), (7) are independent of the shape of the confining region (provided the wavelength is very much less than any gross dimension) proceeds as follows. It is convenient to consider the two dimensional version which can be easily visualised - a membrane stretched with uniform tension and clamped around the boundary.

There are two common boundary conditions for the equation of motion: (1) displacement $u = 0$ on the boundary (2) Normal derivative of displacement $^{\partial u}/\partial n = 0$ on the boundary (the latter is not easy to realise physically).

The first step is to show that $u = 0$ is a generally more restrictive condition than $^{\partial u}/\partial n = 0$: that is, there are more normal modes below a given wave number k for $^{\partial u}/\partial n = 0$, when the edges can go up and down, than for $u = 0$.

Let the number of modes with wave number $<k$ be $N(<k)$ for $u = 0$ on the boundary and $M(<k)$ when $^{\partial u}/\partial n = 0$ on the boundary. Then $N(<k) \lesssim M(<k)$.

Consider the membrane clamped on the boundary, and let the boundary be made up of the edges of little squares. Clamp the membrane further, along the grid defining the little squares. This increases the restriction so the totality of normal modes in the little squares is $n(<k) \leqslant N(<k)$.

Now consider the membrane with the edges unclamped but $\partial u/\partial n = 0$ along the boundary. Cut along the grid lines and constrain the new boundaries by $\partial u/\partial n = 0$. The edges flap in the breeze, the new condition is less restrictive and so the total number of modes $m(<k) \geqslant M(<k)$.

Therefore $\qquad n(<k) \leqslant N(<k) \leqslant m(<k)$

The number of modes of each little square is the same for the two sets of boundary conditions, provided the wavelength is small, and proportional to k^2a where a is the area of each little square. This is the desired result, easily extended to three dimensions, granted the plausible steps listed which of course have to be proved rigorously. (The proof is completed by approximating an arbitrary boundary by a square network from both within and without). The full proof may be found in Courant and Hilbert - Methods of Mathematical Physics I, Ch. VI, Sect. 4. A physical justification due to Peierls is given in Born and Huang - Dynamical theory of Crystal Lattices, Appendix IV. (NB these references are meant for theoretical enthusiasts.)

Given the expression (7) for the density of states we may transform it into any variables we choose to use. In particular

$$\frac{dn}{dE} = \frac{4\pi p^2}{(2\pi\hbar)^3} V \frac{dp}{dE} = \frac{4\pi}{(2\pi\hbar)^3} V\, m\sqrt{2m}\, E^{\frac{1}{2}} \tag{8}$$

The number of particles lying between energy ε and $\varepsilon + \Delta\varepsilon$ is given by

$$\alpha e^{\beta\varepsilon} \frac{dn}{d\varepsilon} \Delta\varepsilon$$

The total number is to be N and their energy E so that we determine α and β via

$$\int_0^\infty \alpha e^{\beta\varepsilon} \frac{4\pi}{(2\pi\hbar)^3} V\, m\sqrt{2m}\, \varepsilon^{\frac{1}{2}} d\varepsilon = N \tag{9}$$

and

$$\int_0^\infty \alpha e^{\beta\varepsilon} \frac{4\pi}{(2\pi\hbar)^3} V m\sqrt{2m}\, \varepsilon^{3/2} d\varepsilon = E \tag{10}$$

where β must be negative for the integrals to converge. We have to evaluate

$$\int_0^\infty e^{\beta\varepsilon}\varepsilon^{\frac{1}{2}}d\varepsilon \qquad \text{and} \qquad \int_0^\infty e^{\beta\varepsilon}\varepsilon^{3/2}d\varepsilon$$

Now
$$\int_0^\infty e^{-kx}x^{\frac{1}{2}}dx = 2\int_0^\infty e^{-ky^2}y^2dy = -2\frac{\partial}{\partial k}\int_0^\infty e^{-ky^2}dy$$

$$\int_0^\infty e^{-kx}x^{3/2}dx = 2\int_0^\infty e^{-ky^2}y^4dy = 2\frac{\partial^2}{\partial k^2}\int_0^\infty e^{-ky^2}dy$$

There is a simple trick for evaluating the last integral:

$$\int_0^\infty e^{-ky^2}dy\int_0^\infty e^{-kx^2}dx = \int_0^\infty e^{-kr^2}\frac{\pi}{2}rdr = \frac{\pi}{4k}$$

Then
$$-2\frac{\partial}{\partial k}\{\tfrac{1}{2}\sqrt{\frac{\pi}{k}}\} = \frac{1}{2}\frac{\sqrt{\pi}}{k^{3/2}}$$

$$2\frac{\partial^2}{\partial k^2}\{\tfrac{1}{2}\sqrt{\frac{\pi}{k}}\} = +\frac{3}{4}\frac{\sqrt{\pi}}{k^{5/2}}$$

Then from (9) and (10)

$$\alpha\frac{4\pi}{(2\pi\hbar)^3}Vm\sqrt{2m}\,\frac{1}{2}\frac{\sqrt{\pi}}{(-\beta)^{3/2}} = N$$

$$\alpha\frac{4\pi}{(2\pi\hbar)^3}Vm\sqrt{2m}\,\frac{3}{4}\frac{\sqrt{\pi}}{(-\beta)^{5/2}} = E$$

So
$$-\frac{3}{2}\frac{1}{\beta} = \bar{E} = \frac{E}{N} : \quad E = -\frac{3}{2}N\frac{1}{\beta}$$

and we may identify $-\beta$ with $^1/kT$.

The number of components with energies between ε and $\varepsilon + \Delta\varepsilon$ is now given by

$$\Delta n = \frac{N\varepsilon^{\frac{1}{2}}e^{-\varepsilon/kT}}{\sqrt{\frac{\pi}{4}}\,(kT)^{3/2}}\Delta\varepsilon \tag{11}$$

which is the correctly normalised Maxwell-Boltzmann distribution for an ideal gas.

Clearly the Maxwell-Boltzmann distribution may be expressed if desired as a function of momentum or of velocity of the gas particles.

PROBLEM: Work out both these forms of the distribution.

Note that the mean energy of a particle is $\frac{3}{2}kT$: when we considered equally spaced component levels in the second lecture we got for the mean energy kT. Clearly the degeneracy of states in our gas has a lot to do with this: at a slightly deeper level we are encountering the equipartition of energy.

PROBLEM: Work out the separation between the single particle energy levels for an ideal gas confined in a box of volume $(10\text{cm})^3$ at $\varepsilon \sim kT$ for room temperature. Estimate the occupation numbers of levels with $\varepsilon \ll kT$ and $\varepsilon \sim kT$.

PROBLEM: Estimate the number of distinct microstates accessible to this gas. Estimate the number of collisions occurring each second in the gas and hence the time it would take for the gas to pass through essentially all the microstates.

PROBLEM: Estimate the separation between adjacent energy states of the WHOLE gas, and hence estimate the time it would take for the gas to settle into a state of well defined energy.

Lecture 6

THE EQUIPARTITION OF ENERGY AND SOME SPECIFIC HEATS

In a system with evenly spaced component energy levels the mean component energy is kT: in an ideal gas the mean component energy is $\frac{3}{2}kT$. We may calculate the mean energy for the ideal gas by a different route:

$$\bar{E} = \frac{\int_0^\infty \varepsilon p^2 dp e^{-\varepsilon/kT}}{\int_0^\infty p^2 dp e^{-\varepsilon/kT}} \quad , \qquad \varepsilon = p^2/2m \tag{1}$$

Because we can write $\varepsilon = \frac{1}{2m}\{p_x^2 + p_y^2 + p_z^2\}$ and because p^2dp is proportional to $dp_x\, dp_y\, dp_z$ we can rewrite (1) in the form

$$\bar{E} = \frac{\frac{1}{2m}\int_0^\infty \{p_x^2 + p_y^2 + p_z^2\} e^{-(p_x^2 + p_y^2 + p_z^2)/2mkT} dp_x\, dp_y\, dp_z}{\int_0^\infty e^{-(p_x^2 + p_y^2 + p_z^2)/2mkT} dp_x\, dp_y\, dp_z}$$

and because the exponentials factor into a product of terms depending only on p_x, p_y or p_z this expression is equivalent to

$$\bar{E} = \frac{1}{2m} \frac{\int_0^\infty p_x^2 e^{-p_x^2/2mkT} dp_x}{\int_0^\infty e^{-p_x^2/2mkT} dp_x} + \text{two more terms in } p_y \text{ and } p_z \tag{2}$$

This at once makes it clear that the mean energy associated with motion in each of the three orthogonal directions is equal, and so equal to $\frac{1}{2}kT$, a result which may be checked directly from Eq. (2). It should also be clear that the one and two dimensional versions of this problem will yield results for $\bar{E}$ equal to $\frac{1}{2}kT$ and kT respectively. That is, $\frac{1}{2}kT$ is the mean energy associated with each of the three degrees of freedom of each particle (freedom to move in the x, y and z directions independently).

PROBLEM: Show that the energy levels of a particle in a two dimensional box are uniformly spaced (on average) in energy.

PROBLEM: Consider a stretched string in thermal contact with a heat bath. What is its mean vibrational energy? What do the occupation numbers mean for such a system?

Consider a different sort of component with uniformly spaced energy levels: the (one dimensional) simple harmonic oscillator with a Hamiltonian

$$H = \frac{p_x^{\,2}}{2m} + \frac{1}{2}kX^2, \quad \text{frequency } \omega.$$

The allowed energies are $\varepsilon = (n + \frac{1}{2})\hbar\omega$ and p, x are now <u>internal</u> quantities, for example, momenta and coordinates of vibrations of a diatomic molecule introduced into our ideal gas. Because the levels are equally spaced the mean energy is kT, even though the problem is one dimensional. But remember that in a harmonic oscillator the mean kinetic energy is equal to the mean potential energy. The kinetic energy has a mean value $\frac{1}{2}kT$, just as for one degree of freedom of a particle in a box, and the potential energy carries another $\frac{1}{2}kT$. When the energy is additive in $p_x^{\,2}$, $p_y^{\,2}$, $p_z^{\,2}$; x^2, y^2, z^2 each of these terms if present carries $\frac{1}{2}kT$ on average, provided kT is large in comparison with the level spacing. A degree of freedom is associated with each of the <u>coordinates in phase space</u> p_x, p_y, p_z, x,y,z which appears quadratically in the Hamiltonian.

We can easily calculate exactly the mean energy of a one dimensional harmonic oscillator:

$$\bar{E} = \frac{\sum_0^\infty (n + \frac{1}{2})\hbar\omega e^{-(n + \frac{1}{2})\hbar\omega/kT}}{\sum_0^\infty e^{-(n + \frac{1}{2})\hbar\omega/kT}} = \frac{1}{2}\hbar\omega + \hbar\omega\frac{\sum_0^\infty n e^{-n\hbar\omega/kT}}{\sum_0^\infty e^{-n\hbar\omega/kT}} \qquad (3)$$

The first term does not depend on temperature and is the zero point energy of the oscillator. The second term may be evaluated as follows:-

$$(1 - x)^{-1} = 1 + x + x^2 \ldots = \sum_0^\infty x^n$$

$$(1 - x)^{-2} = 1 + 2x + 3x^2 \ldots$$

$$x(1 - x)^{-2} = x + 2x^2 + 3x^2 \ldots = \sum_0^\infty nx^n$$

so $$\frac{\Sigma nx^n}{\Sigma x^n} = \frac{x}{1-x} \quad : \text{ set } x = e^{-h\omega/kT} \qquad \text{when}$$

$$\overline{E} = \frac{1}{2}\hbar\omega + \frac{\hbar\omega e^{-\hbar\omega kT}}{1 - e^{-\hbar\omega/kT}} = \frac{1}{2}\hbar\omega\left\{\frac{1 + e^{-\hbar\omega/kT}}{1 - e^{-\hbar\omega/kT}}\right\}$$

$$\rightarrow kT \text{ for } kT >> \hbar\omega$$

A three dimensional harmonic oscillator will have 6 degrees of freedom and hence mean energy 3kT.

In passing, let us construct a stupid model for the specific heat of a solid. The solid contains N atoms which vibrate about their equilibrium positions. Let us suppose (this is the stupid bit) that they only interact weakly and so can be regarded as N independent three dimensional harmonic oscillators. The total number of degrees of freedom is 6N and so the mean energy is 3NkT - if N_A is the number in a mole, $C_V = 3N_A k = 3R$, which is the Dulong and Petit result for the specific heat of a metal.

[PROBLEM: Why C_V?]

On the other hand, at low temperatures $\bar{E} \sim \frac{1}{2}\hbar\omega + \hbar\omega e^{-\hbar\omega/kT}$ for our one dimensional harmonic oscillator so

$$C_V \simeq 3N_A\hbar\omega\frac{\partial}{\partial T}e^{-\hbar\omega/kT} = 3R\left(\frac{\hbar\omega}{kT}\right)^2 e^{-\hbar\omega/kT} \qquad (4)$$

which is roughly right but not quite right. The model is maybe not so stupid - it is Einstein's model of the specific heat of solids. (Read about Debye's improved model in, for example, Kittel's Introduction to Solid State Physics.)

We have dealt with vibrational states of molecules: what about rotational states? The energy of a rotating rod is $\frac{1}{2}\frac{\Omega^2}{I}$, where Ω is the angular momentum and I is the moment of inertia. In quantum mechanics

$$\Omega^2 \rightarrow \hbar^2 J(J + 1)$$

where J is an integer and a state with angular momentum J is (2J +1) degenerate (because of the magnetic quantum number). Therefore, setting $\frac{1}{2}\hbar^2/I = \alpha$, we have

$$\bar{E} = \frac{\Sigma\alpha(2J + 1)J(J + 1)e^{-\alpha J(J + 1)/kT}}{\Sigma(2J + 1)^{-\alpha J(J + 1)/kT}}$$

and at high temperatures ($kT \gg \alpha$)

$$\bar{E} \sim \frac{\int \alpha J^3 e^{-\alpha J^2/kT} dJ}{\int Je^{-\alpha J^2 kT} dJ} = kT$$

because the rotating rigid rod has two degrees of freedom. (If you leave out the degeneracy factor (2J + 1) you only get $\frac{1}{2}kT$.)

We may now consider the specific heats of gases. Consider our ideal gas confined within a box. Supply energy to this gas but allow no change to the volume of the box. The energy levels are unchanged and the energy supplied is rapidly distributed so that the gas passes to a new equilibrium configuration with a higher mean energy - which means a higher temperature. We have supplied HEAT to the gas.

For a mole of perfect monatomic gas, the total energy is

$$E = \bar{E}N_A = \frac{3}{2}N_A kT = \frac{3}{2}RT$$

(where N_A is Avogadro's number). If the gas is diatomic but perfect in other respects, we will find an extra kT from the vibrational degrees of freedom and an extra kT from the rotational degrees of freedom. Thus for a diatomic gas

$$\bar{E} = \frac{7}{2}kT$$

Now the vibrational and rotational energy levels are (relatively) widely separated in comparison with the kinetic energy levels, so as the temperature drops we expect $\bar{E} = \frac{7}{2}kT \rightarrow \frac{3}{2}kT$ for a diatomic gas, as kT drops below the various interlevel spacings. Clearly the specific heat at constant volume is given by

$$C_V = \left(\frac{\partial E}{\partial T}\right)_V = \frac{1}{2}nR \tag{5}$$

where n is the number of active (classical) degrees of freedom. This sort of behaviour is exactly what may be observed and it caused consternation in the latter half of the last century, because classically all energy levels are blurred into the appropriate continuum, the classical treatment then being valid for any temperature, and classically degrees of freedom NEVER freeze out. So the specific heats of gases were the first signpost to quantum theory!

Now what about the specific heat of an ideal gas at constant pressure as opposed to constant volume? Why is this different, and can we calculate the difference? Obviously the volume of the container changes, but the value of $\bar{E}$ for the kinetic energy of a single molecule didn't depend on the volume of the box. But suppose we shove energy into a gas in a box which is connected to a mercury manometer. At constant volume molecules pick up energy, although the energy levels remain the same, and that is all. At constant pressure the molecules pick up energy, collide more vigorously with the mercury and push it up the manometer tube. The volume increases, the energy levels change, and some of the energy we have shoved into the gas is used up lifting the mercury against gravity. We therefore have to put in more energy in order to raise the gas temperature through a specified interval at constant pressure than at constant volume.

Now we consider the mechanics of molecules bouncing off the walls of the container. (They are in thermal equilibrium with the walls so it doesn't matter if they stick for a bit and then come off with various velocities and directions: on average we can treat the wall as a mirror.) The old kinetic theory argument, perfectly sound for equilibrium, calculates the change of momentum of a molecule in the process

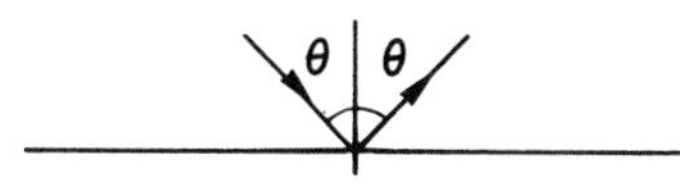

Momentum transferred to the wall by 1 molecule = 2 $mv\cos\theta$. Let $n(v,\theta)$ be the density of molecules with velocity (v,θ). Then the momentum transferred by molecules of velocity (v,θ) is given by $2mv^2n(v,\theta)\cos^2\theta$ normal to the wall, each second, per unit area (and solid angle). This is because the number of molecules striking unit area is $n(v,\theta)v\cos\theta$ each second, and each yields up momentum $2mv\cos\theta$. Now the number of molecules between θ and $\theta + d\theta$ is $n(v,\theta)\,2\pi d\cos\theta$

and $$\int n(v,\theta)2\pi d\cos\theta = n(v) = 4\pi n(v,\theta)$$

because $n(v,\theta)$ is independent of θ.

So the momentum transferred to unit area of the walls each second, which is the pressure, is given by

$$P = \int_0^\infty\int_0^1 2mv^2 n(v,\theta)\cos^2\theta 2\pi d\cos\theta dv$$

$$= \frac{4\pi}{3}\int mv^2 n(v,\theta)dv$$

$$= \frac{1}{3}\int mv^2 n(v)dv$$

$$= \frac{2}{3}\int\frac{1}{2}mv^2 n(v)dv = \frac{2}{3}\;\frac{1}{2}\;\overline{mv^2 n}$$

$$= \frac{2}{3}\;\bar{E}_{KE}n$$

where n is the number density. For 1 mole of gas

$$P = \frac{2}{3}\;\bar{E}_{KE}\;\frac{N_A}{V}$$

so $$PV = \frac{2}{3}\;\bar{E}_{KE}\;N_A = N_A kT = RT \qquad (6)$$

of course the equation of state of an ideal gas.

The work done at constant pressure by the gas on the outside world is $P\Delta V$. We are changing the temperature so

$$\Delta V = \frac{R\Delta T}{P}$$

and the work done is therefore $R\Delta T$. Then for an ideal gas

$$C_p - C_v = R \qquad (7)$$

for if an amount of energy E is put in

$$\left(\frac{\partial E}{\partial T}\right)_P = R + \left(\frac{\partial E}{\partial T}\right)_V$$

and for a monatomic gas, $C_V = \frac{3}{2}R$, $C_P/C_V = \frac{5}{3}$.

So long as the kinetic degrees of freedom are behaving ideally, and the gas density is sufficiently low that the rotational and vibrational energy states are those of a free molecule, the result (7) may be extended to molecular gases, the internal degrees of freedom coming in C_V. Thus for a rotating, non-vibrating diatomic gas $C_v = \frac{5}{2}R$ and $^{C_p}/_{C_v} = \frac{7}{5}$ while if both rotational and vibrational levels are excited, $^{C_p}/_{C_v} = \frac{9}{7}$.

[PROBLEM: Why don't we have to worry about inelastic collisions of complex molecules with the walls?]

Note that as soon as we let the volume of the container change we had to consider external constraints and their interaction with the system.

Lecture 7

SOME THERMODYNAMIC FUNCTIONS

In the previous lecture we calculated the pressure of an ideal gas by using arguments from kinetic theory and found the difference between C_P and C_V by calculating the work done by the gas on the outside world at constant pressure as the gas expanded slightly: (pressure = force per unit area) x (V = area x distance moved) = work done.

We may calculate the pressure by an entirely different technique, using the idea of virtual work. The change of energy in the system for an infinitesimal expansion is the work done, so we can identify

$$P = -\left(\frac{\partial E}{\partial V}\right)_X$$

where the X represents anything kept constant in the process. We are taking the limit where the energy does not in fact change and the occupation numbers are in equilibrium, so the appropriate thing to do is to keep the occupation numbers constant, and

$$P = -\left(\frac{\partial E}{\partial V}\right)_{n_i} \qquad (1)$$

(The volume increase is done very very slowly so the levels sag but the populations remain the same.) Now $E = \Sigma\varepsilon_i n_i$

So $$\Delta E = \Sigma n_i \Delta\varepsilon_i + \Sigma\varepsilon_i \Delta n_i \qquad (2)$$

For constant volume, an increase of internal energy changes the occupation numbers n_i, but no work is done on the system or by the system: the second term represents a change of HEAT. If the volume is infinitesimally changed the first effect is a minute change in the energy levels ε_i: the first term represents work done on the system. We can thus write

$$dE = đW + đQ \tag{3}$$

which is the first law of thermodynamics. Note the bars on the differentials on the right hand side of (3): these are to indicate that $đW$ and $đQ$ are infinitesimal changes but are not exact differentials. (The quantity $d(xy) = xdy + ydx$ is an exact differential: xdy by itself may not be, and in general it is not.) The distinction will become clear later.

Thus we can write

$$P = -\Sigma\, n_i \frac{\partial \varepsilon_i}{\partial V} \tag{4}$$

There are various ways of working out this expression. We have

$$n_i = \frac{Ne^{-\varepsilon_i/kT}}{\Sigma e^{-\varepsilon_i/kT}} = \frac{Ne^{-\varepsilon_i/kT}}{z}$$

and the simplest and most useful thing is to note that

$$\frac{\partial}{\partial V}\{\ell n z\}_T = \frac{1}{z}\left(\frac{\partial z}{\partial V}\right)_T$$

$$= \frac{-\Sigma \frac{\partial \varepsilon_i}{\partial V} \frac{1}{kT} e^{-\varepsilon_i/kT}}{z}$$

so that $$P = NkT\frac{\partial}{\partial V}\{\ell n z\}_T \tag{5}$$

For our gas, $z = \int e^{-\varepsilon/kT} \frac{4\pi p^2 dp}{(2\pi\hbar)^3} V$

so that $\frac{\partial}{\partial V}\{\ell n z\}_T = \frac{1}{V}$

and of course we recover the equation of state of an ideal gas,

$$PV = NkT$$

It is convenient to define an energy function

$$F = -NkT\ell n z, \quad z^N = e^{-F/kT} \tag{6}$$

when $$P = -\left(\frac{\partial F}{\partial V}\right)_T \tag{7}$$

This quantity is the Helmholtz free energy (it will require modification later on when we worry about indistinguishability of particles in a gas) and is such that the differential with respect to volume, at constant temperature, gives the pressure. (The differential of the internal energy E at constant temperature does NOT give the pressure.) The free energy is thus the work we can in principle extract from our gas if it is maintained at constant temperature.

Note that the internal energy can also be expressed in terms of the partition function

$$E = \frac{N\Sigma\varepsilon_i e^{-\varepsilon i/kT}}{z} = NkT^2 \frac{\partial \ell nz}{\partial T} \tag{8}$$

(the latter form can save us the trouble of evaluating sums or integrals twice).

Now consider what we can control or observe from outside our box of gas: the temperature, the volume and the pressure. We have

$$-\left(\frac{\partial F}{\partial V}\right)_T = NkT\left(\frac{\partial \ell nz}{\partial V}\right)_T = P \tag{9}$$

and work done ON the system by the outside world is

$$đW = -PdV = \left(\frac{\partial F}{\partial V}\right)_T dV$$

The equation of state links P, V and T and so we may write

$$dF = \left(\frac{\partial F}{\partial V}\right)_T dV + \left(\frac{\partial F}{\partial T}\right)_V dT$$

so that

$$-PdV = dF - \left(\frac{\partial F}{\partial T}\right)_V dT$$

and

$$\left(\frac{\partial F}{\partial T}\right)_V = -Nk\ell nz - \frac{NkT}{z}\frac{\partial z}{\partial T} \tag{10}$$

Now if W is the number of microstates accessible to the system, remember that

$$\ell nW \simeq N\ell nN + \Sigma n_i \ell ng_i - \Sigma n_i \ell nn_i$$

where g_i is the degeneracy of the states (see Lecture 4). On substituting $n_i = Ng_i e^{-\varepsilon i/kT}/z$ this expression becomes

$$\ell nW = \frac{E}{kT} + N\ell nz \tag{11}$$

so that we can write

$$E = -NkT\ell nz + kT\ell nW = F + kT\ell nW \qquad (12)$$

and

$$dE = dF + (k\ell nW)dT + kTd(\ell nW)$$
$$= -PdV + \left(\frac{\partial F}{\partial T}\right)_V dT + (k\ell nW)dT + kTd(\ell nW) \qquad (13)$$

Now $\left(\frac{\partial F}{\partial T}\right)_V = -Nk\ell nz - \frac{NkT}{z}\frac{\partial z}{\partial T} = -Nk\ell nz - \frac{E}{T}$ (14)

So $\left(\frac{\partial F}{\partial T}\right)_V = -k\ell nW$ (15)

$E = F + kT\ell nW$ (from 12)

The quantity $d(\ell nW)$ is a perfect differential. Give $k\ell nW$ a name - S - and write from (12)

$$E = F + TS \qquad (12)$$

and using (15) in (13) we also have

$$dE = -PdV + TdS \qquad (16)$$

The results (12) and (16), seen side by side, look a bit peculiar. Differentiate (12) and get

$$dE = dF + TdS + SdT$$

This expression reduces to (13) because the term in SdT is cancelled by that part of dF which does not correspond to work.

We may now write

$đW = -PdV$ (work done ON the system)

$đQ = TdS$

and we have found a quantity, $S = k\ell nW$, which we may identify with the thermodynamic function known as the entropy. (We shall have to modify some details when we worry about indistinguishability of identical particles in a gas).

We may notice that once we have specified (for our isolated system) the mass of gas, everything else is given by specifying the energy levels and the distribution of particles over them (that is, the volume and temperature). If alternatively we were considering a set of spins in a magnetic field, we would specify the number of spins, the magnetic field (which would specify the available energy levels) and the temperature. We can clearly envisage situations where the component energy levels depend on more than one external parameter.

Once our system is specified by giving its macroscopic properties, then the internal energy E, the free energy F and entropy S have unique values: they are thermodynamic functions of state (and can all be calculated from z). It should be clear however that we could prepare a specified system in a number of different ways from given initial conditions. We might change the volume first, then heat the gas, or we might heat it first and then change the volume; the volume might be changed adiabatically (no heat allowed in) or isothermally. In general the amount of work done and the amount of heat added to reach a specified state depend on how we do it and so in general we cannot talk about a well defined work content or heat content of a system (although this does not stop you talking loosely about the heat content of a block of copper). This is the physical reason why $đW$ and $đQ$ are not perfect differentials.

Suppose we consider two or more systems which may or may not be in thermal contact. The energy is clearly additive. The number of possible microstates of the whole set of systems is clearly the product of the number of possible microstates of the individual systems

$$W = \prod_i w_i \qquad \therefore \qquad \ln W = \Sigma \ln w_i$$

and so the <u>entropy is an additive heat quantity</u>. Since E = F + TS, F is additive only if the systems are in thermal equilibrium.

Quantities which do not change when you increase a number of identical systems by 1 are called intensive quantities, while a quantity which is proportional to the number of identical systems is called an extensive quantity. Thus P, T are intensive, while

V, E, S, F are extensive

Some features which will be useful later are:

1. We may define the temperature of a single component to be the temperature of a macroscopic system of which it is part; or with which it is in thermal contact.

2. The contribution of a single particle to the entropy is

$$\frac{S}{N} = k\ell nz + \frac{\bar{E}}{T}$$

$$= k\ell nz + \frac{1}{T}\frac{\Sigma\varepsilon_i e^{-\varepsilon_i/kT}}{z}$$

$$= k\ell nz - k\frac{\Sigma e^{-\varepsilon_i/kT}}{z}\ell n(e^{-\varepsilon_i/kT})$$

$$= -k\frac{\Sigma e^{-\varepsilon_i/kT}}{z}\ell n\left(\frac{e^{-\varepsilon_i/kT}}{z}\right)$$

$$= -k\Sigma P_i \ell n P_i$$

where P_i is the probability that a given particle is in the ith state (see the fourth lecture). (Incidentally, here we make contact with information theory.)

(The entropy is only related to E, F and to the partition function in the way we have worked out for states of thermodynamic equilibrium. Definitions in terms of the number of accessible states or probability can be used more generally.)

3. Similarly, the contribution of a single particle to F is

$$\frac{F}{N} = -kT\ell nz$$

The utility of a definition of pressure in terms of an appropriate derivative of the internal energy is that we may at once introduce a generalised pressure. Let the energy levels depend on a set of parameters x_i: a generalised pressure term is then

$$-\left(\frac{\partial E}{\partial x_i}\right)_n$$

and we can define a generalised work term

$$đW = +\Sigma\left(\frac{\partial F}{\partial x_i}\right)_T dX_i$$

For example, if we consider spins in an external magnetic field B,

$$\varepsilon_i = -\underline{\mu}.\underline{B} \text{ which we write as } -\mu_i B$$

so
$$E = \frac{N\Sigma(-\mu_i B)e^{+\mu_i B/kT}}{\Sigma e^{+\mu_i B/kT}}$$

and
$$\left(\frac{\partial E}{\partial B}\right)_n = \frac{N\Sigma(-\mu_i)e^{+\mu_i B/kT}}{\Sigma e^{+\mu_i B/kT}} = -M$$

and the work done on the system by a small change in the field B is given by

$$đW = -MdB$$

However, BEWARE of a facile identification of this term with the work done magnetising a specimen in a solenoid, because the physics of the interaction of the spins with the outside world has to be thought out carefully. See for example Kittel Thermal Physics, Ch. 23, Mandl Statistical Physics, Ch. 1.4. Understanding this material constitutes an entirely adequate PROBLEM for this lecture!

Lecture 8

THERMODYNAMICS OF AN IDEAL GAS

Although the last lecture was concerned primarily with an ideal gas, the definitions of entropy S and free energy F generalise straightforwardly to other systems of weakly interacting components in equilibrium. The entropy S is of fundamental importance and is given by $S = k\ln W$ where W is the total number of microstates available to the system.

For a system in equilibrium the internal energy, free energy and the entropy are related by

$$E = F + TS \tag{1}$$

and we can also write

$$dE = TdS - PdV \tag{2}$$

for the specific case of a gas. Note that in calculating the differentials we assumed equilibrium configurations and equation (2) thus only applies to changes in which equilibrium is maintained: that is, external constraints are changed on a time scale large in comparison with the relaxation time of the system. Such processes are called reversible processes, and we may consider as an example the Carnot cycle for an ideal gas.

Suppose we start by letting a gas expand at constant temperature - that is, in thermal contact with a heat reservoir. The gas does work

$$đW_{\uparrow} = PdV$$

in pushing back a piston*. The expansion is isothermal and we have already established the equation of state

---

* There must be a piston because the process must be slow.

$$PV = NkT$$

so P is a function of V only and we may integrate the expression for the work done BY the gas

$$\Delta W_{\uparrow} = \int_{V_1}^{V_2} PdV = NkT \int_{V_1}^{V_2} \frac{dV}{V} = NkT\ell n\left(\frac{V_2}{V_1}\right) \qquad (3)$$

Now $$dF = \left(\frac{\partial F}{\partial V}\right)_T dV + \left(\frac{\partial F}{\partial T}\right)_V dT$$

and the expansion is isothermal so $\Delta F = -\Delta W_{\uparrow}$ where I am using an ↑ to indicate the gas doing work on the outside world.

For a gas at a temperature T, the internal energy is

$$E = \frac{3}{2}NkT$$

and this expression is independent of the volume or the pressure. Then in this isothermal process the internal energy remains constant and

$$dE = 0 = dQ - PdV \qquad (4)$$

It is clear that an amount of heat

$$\Delta Q = NkT\ell n\left(\frac{V_2}{V_1}\right)$$

must flow from the reservoir at temperature T into the gas. We have

$$đQ = TdS \qquad (5)$$

and the process is isothermal so we can integrate (5) and write

$$\Delta Q = T\Delta S = NkT\ell n\left(\frac{V_2}{V_1}\right)$$

whence $$\Delta S = Nk\ell n\left(\frac{V_2}{V_1}\right)$$ in this process.

Let's check this explicitly from the expression we obtained in the last lecture for entropy of an ideal gas in equilibrium

$$S = Nk\ell nz + \frac{E}{T}$$

The difference in S between the two equilibrium configurations is

$$\Delta S = Nk\ell n\frac{z_2}{z_1}$$

and remember that $z = \int \frac{dn}{d\varepsilon} e^{-\varepsilon/kT} d\varepsilon \propto V$ so that for the same initial and final values of the temperature

$$\Delta S = Nk\ell n\frac{V_2}{V_1}$$

once more.

We expect the entropy to have increased, because the internal energy has remained constant and the energy levels have got closer together.

Now isolate the system from any heat reservoir, and allow it to expand adiabatically from temperature T to temperature T'. No heat can flow into the system but it can still do work, drawing on the internal energy: we therefore expect the temperature to drop.

If no heat flows into the gas, then at every small step

$$đQ = TdS = 0$$

and so this adiabatic expansion should involve no change of entropy. Thus

$$S = Nk\ell nz + \frac{E}{T} = Nk\ell nz + \frac{3}{2}Nk$$

should remain constant as the system evolves through equilibrium configurations. The particle partition function z depends linearly on V and also depends on T

$$z = \int \frac{dn}{d\varepsilon} e^{-\varepsilon/kT} d\varepsilon$$

Since $\frac{dn}{dE} \propto Vp^2\frac{dp}{dE} \propto VE^{\frac{1}{2}}$, $z \propto VT^{3/2}$

so the constant entropy condition requires

$$VT^{3/2} = \text{constant} \tag{6}$$

Let the expansion go from V_2 to V_3 while the temperature goes from T to T'. Then

$$V_2T^{3/2} = V_3T'^{3/2} \; ; \; T' = \left(\frac{V_2}{V_3}\right)^{2/3} T < T \tag{6a}$$

The decrease in internal energy all goes in work, so the work done in this leg of the cycle is

$$\Delta W_{\uparrow} = \frac{3}{2}Nk(T - T')$$

where T' is given by (6). [We ought to be able to work this out without even mentioning entropy. Simply require that the internal energy decreases by an amount equal to the work done, given

$$PV = NkT \text{ and } E = \frac{3}{2}NkT$$

First step. Let the volume expand by ΔV. Then $\Delta W_{\uparrow} = P\Delta V = \frac{NkT}{V}\Delta V$

Second step. To calculate ΔT, set

$$\Delta E = \frac{3}{2}Nk\Delta T = -\Delta W_{\uparrow} = -\frac{NkT}{V}\Delta V$$

so that
$$\frac{\Delta T}{T} = -\frac{2}{3}\frac{\Delta V}{V}$$

which yields (6a). We have a direct check that we didn't do anything stupid working with the entropy.]

In the adiabatic process, the entropy remains constant during the expansion because although the energy levels are getting closer together, the internal energy is decreasing so as to keep the number of microstates accessible to the system the same.

Now we can confidently go on to complete the rest of our Carnot cycle. The gas is at a temperature T' as a result of the adiabatic expansion. Bring it into thermal contact with a reservoir at this temperature T' and compress the gas. Work is done ON the gas but the temperature and hence the internal energy remain fixed. Therefore heat flows INTO the reservoir at temperature T'.

The work done ON the gas is

$$NkT'\ln\left(\frac{V_3}{V_4}\right) \qquad V_4 < V_3$$

and an equal amount of heat energy flows out of the gas into the reservoir.

Choose V_4 such that on recompressing adiabatically from V_4 to V_1 we return to the original state of the system. This condition requires that

$$\frac{T}{T'} = \left(\frac{V_4}{V_1}\right)^{2/3} \quad ; \quad V_1 < V_4 \tag{7}$$

so that
$$\frac{V_4}{V_1} = \frac{V_3}{V_2}$$

The work done ON the gas is

$$\Delta W_{\downarrow} = \frac{3}{2}Nk(T - T')$$

(where I use $\downarrow$ to denote work done on the gas) and so we get no net work out of the adiabatic parts of the cycle.

The total work done BY the gas ON the piston is

$$W_{\uparrow} = NkT\ln\left(\frac{V_2}{V_1}\right) - NkT'\ln\left(\frac{V_3}{V_4}\right)$$

and this is done during the isothermal legs. The amount of energy extracted as heat from the reservoir at temperature T is

$$Q_T = NkT\ell n(\frac{V_2}{V_1})$$

So the proportion of heat energy extracted from the high temperature reservoir which appears as work is given by

$$\frac{T\ell n(\frac{V_2}{V_1}) - T'\ell n(\frac{V_3}{V_4})}{T\ell n(\frac{V_2}{V_1})} = 1 - \frac{T'}{T}$$

because $\frac{V_3}{V_4} = \frac{V_2}{V_1}$, from (7).

Note that, as a result of this condition, adding up the changes of entropy of the gas in each leg of the cycle gives a total entropy change of zero, as it should.

We have calculated the thermodynamic efficiency of a Carnot cycle operating between a reservoir at temperature T and another at temperature T'. This efficiency only approaches 1 as $T'/T \to 0$. The reversed Carnot cycle (isothermal expansion at the lower temperature followed by isothermal compression at the higher temperature) requires the input of work - the heat pump cycle as opposed to the heat engine.

In the four legs of the Carnot cycle discussed above, we moved slowly through a sequence of equilibrium states and consequently we could follow the development of the system continuously. Such very slow processes are reversible (in general visualise an arbitrary very slow process as consisting of a sequence of isothermal and adiabatic legs) in the sense that any work got out is just sufficient to drive the reverse process.

[PROBLEM: Demonstrate explicitly that the whole Carnot cycle is reversible.]

If however we make a rapid change, the external constraints are changed in a time short in comparison with the relaxation time of the system and we cannot follow its evolution using equilibrium statistical mechanics. We can however compare the initial and final states.

Suppose for example we have a box of volume V_2 and the gas is confined within $V_1 < V_2$ by a partition. If the partition is suddenly withdrawn the gas expands into the full volume but can do no work on the outside world. If it is an ideal gas, and our calculations have been for an ideal gas, $E = \frac{3}{2}NkT$ and so both the internal energy and the temperature remain the same. The energy levels are now closer together and so the entropy has increased

$$\Delta S = Nk\ell n(\frac{V_2}{V_1})$$

so in this process $T\Delta S > \Delta Q = 0$. It is clear that to restore the system to its original state we either wait until the gas happens to be all back in the subsidiary volume V_1 and then slam the door - and this NEVER happens for a macroscopic system - or we must compress the gas into V_1 which requires work to be done against the pressure. We didn't get any work out of the original expansion. Furthermore, in the expansion the entropy of a closed system increased. At every stage of the reversible Carnot cycle, the entropy of the whole system (two reservoirs + heat engine) remained constant. These fast processes are irreversible (which doesn't mean that you can't get back to the original setup, only that you have to do work you didn't get out in the first instance.)

We have now established the statistical thermodynamics of a system of weakly interacting components which may or may not be in thermal contact with other systems. We haven't yet considered mixing of different systems .....

Lecture 9

ENTROPY OF MIXING: GIBBS' PARADOX

We have now reached a point where we have used statistical mechanics to show that a system of weakly interacting components, isolated or in thermal contact with a reservoir, obeys the laws of thermodynamics, and we have a prescription for calculating the thermodynamic functions E, S, F in terms of the properties of the components. One example of the kind of system we have been considering is a dilute paramagnetic substance in a magnetic field; another is an ideal gas.

The entropy for a weakly interacting system is given by

$$S = k\ln W \qquad W = N!\,\Pi\,\frac{g_i^{n_i}}{n_i!} \tag{1}$$

whence

$$S = Nk\ln z + {}^{E}/T \tag{2}$$

$$= Nk\ln z + \tfrac{3}{2}Nk \qquad \text{for a gas}$$

where the single component partition function is

$$z = \Sigma e^{-\varepsilon_i/kT}$$

For an ideal gas

$$z \to \int \frac{dn}{d\varepsilon} e^{-\varepsilon/kT} d\,\varepsilon \tag{3}$$

where $$dn = \frac{4\pi p^2 dpV}{(2\pi\hbar)^3}$$

and hence $$\frac{dn}{d\varepsilon} = \frac{4\pi\ V}{(2\pi\hbar)^3}\, m\sqrt{2m}\,\varepsilon^{\frac{1}{2}}$$ (see Lecture 5)

Integrating (3) we obtain

$$z = \frac{4\pi\ V}{(2\pi\hbar)^3} m\sqrt{2m}\ \tfrac{1}{2}\sqrt{\pi}\,(kT)^{3/2} \tag{4}$$

[PROBLEM: Calculate a numerical value for the partition function for Helium gas at room temperature (take a volume of 1 litre). Then calculate a numerical value for the entropy using eq. (2).]

We can therefore write

$$S = Nk[\ell nV + \tfrac{3}{2}\ell nkT + \text{constants}] \tag{5}$$

Now consider two boxes of the same gas at the same temperature, one containing N_1 molecules in volume V_1, the other containing N_2 molecules in volume V_2. The total entropy is

$$S_1 + S_2 = N_1k[\ell nV_1 + \tfrac{3}{2}\ell nkT] + N_2k[\ell nV_2 + \tfrac{3}{2}\ell nkT] + \text{constants } k[N_1 + N_2] \tag{6}$$

Open a partition between the two gases (without doing any work) and wait until a new equilibrium has been established. Then for the entropy of the mixed gas eq. (5) gives us

$$S_{12} = (N_1 + N_2)k[\ell n(V_1 + V_2) + \tfrac{3}{2}\ell nkT + \text{constants}] \tag{7}$$

and this expression is the sum of the entropies of N_1 particles inhabiting a volume $V_1 + V_2$ and of N_2 particles inhabiting a volume $V_1 + V_2$.

Then $S_{12} - (S_1 + S_2) = k\{(N_1 + N_2)\ell n(V_1 + V_2) - N_1\ell nV_1 - N_2\ell nV_2\}$

If $N_1 = N_2 = N$ and $V_1 = V_2$ then

$$S_{12} - (S_1 + S_2) = 2Nk\ell n2 \tag{8}$$

At first sight this seems fine. The molecules originally in the two boxes have had more microstates made available to them and so there has been an increase in entropy.

But now slide a partition back into place. This can be done without doing any work, and it can be done as slowly as we like, so the process is reversible (although the original process may not have been). The pressure is now equalised throughout the volume $V_1 + V_2$, so sliding in the partition gives N_1' in V_1, N_2' in V_2, such that ${}^{N_1'}/V_1 = {}^{N_2'}/V_2$. If we use eq. (5) for the entropy, then we calculate a decrease in entropy on sliding in the partition. This is not only contrary to common sense, but it is also contrary to the thermodynamic equations we worked out in Lecture 8: the process is reversible, adiabatic and no work is done. There should be no entropy change. We have some kind of a contradiction here, and it is serious because if entropy depends on how often we pull a partition in and out, then it cannot be a function of state! This is the first time we have considered mixing.

We would be quite happy to find an increase of entropy if two volumes of two different gases mixed: the problem must be concerned with the mixing of two volumes of the same gas (the components being identical). We will try and resolve the problem by supposing the molecules of a homogeneous gas to be indistinguishable (we are not allowing different excited states for the moment). Just what we mean by indistinguishable has to be looked into rather carefully later on.

In working out the number of microstates corresponding to a given macrostate we counted up the number of ways of getting n_1 in level 1, n_2 in level 2 and so on. For example, if we had two levels and three particles were to be distributed over them with an energy restriction such that two had to be in the lower state and one in the upper state, then we counted

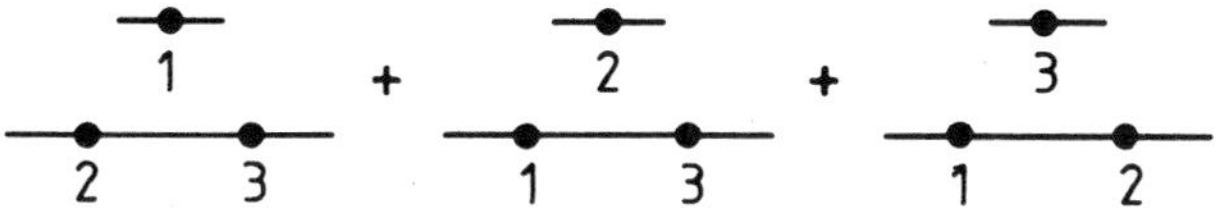

1 + 2 + 3
2 3 1 3 1 2 = 3 microstates

(for this case $N!\Pi \frac{1}{n_i!} = 3!\frac{1}{2!}\frac{1}{1!} = 3$ again).
BUT if all three particles are indistinguishable, then perhaps this means that the macrostate is completely specified by saying there are two particles in the lower level and one in the upper level. If this is the case there is only one microstate in the above example, AND THE SAME IS TRUE OF ANY MACROSTATE. If a macrostate is specified by saying there are n_1 in level 1 ... n_i in level i and the components are indistinguishable then there is only one way of making the macrostate: namely n_1 in 1 ... n_i in i.

Then $W = \frac{N!}{\Pi n_i!}$ must be replaced by 1 and to find the number of accessible states we cannot use Stirling's theorem and calculate the quantity W_{max}.

There is a way out of this difficulty. Suppose that we have a system where the occupation numbers of each level are on average <<1 and momentarily are either 1 or zero (mostly zero). Group the levels together in clumps of g_i approximately degenerate levels and calculate the number of ways of achieving m_i in clump i, etc. We calculated this before (see Lectures 3 and 4) and found

$$W(g_i) = N!\Pi \frac{(g_i)^{m_i}}{m_i!} \tag{9}$$

and $W(g_i)_{max} \simeq \sum_{\text{all macrostates}} W$

With the numbers in each level usually 1 or zero (so $n_i! = 1$), the effect of indistinguishability on this function can be taken into account by dividing $W(g_i)$ by the number of ways of permuting all N components, namely N!.

So for a dilute system we may suppose that with indistinguishable components the total number of states available to the system is given by

$$\bar{W}(g_i) = \Pi \frac{(g_i)^{m_i}}{m_i!} \tag{10}$$

and we may now use Stirling's theorem to maximise this expression, obtaining of course

$$m_i = \frac{Ng_i e^{-\varepsilon_i/kT}}{\Sigma g_i e^{-\varepsilon_i/kT}} = \frac{Ng_i e^{-\varepsilon_i/kT}}{z}$$

Now $\bar{S} = k\ell n\bar{W} = k\{-N\ell nN + \frac{E}{kT} + N\ell nz + N\}$

(that is, $\bar{S} = S - k\ell nN!)$

$$= \frac{E}{T} + Nk\ell n(\frac{z}{N}) + Nk \tag{11}$$

$$\bar{S} = \frac{5}{2}Nk + Nk\ell n(\frac{z}{N})$$

for a gas where as before z is given by eq. (4).

Equation (11) is the Sakur-Tetrode equation for the entropy of an ideal gas and is experimentally correct (see Kittel - Thermal Physics, Ch. 11) and resolves our problems about entropy. It differs from our original expression primarily in that V, the volume, is replaced by $^V/N$, the volume per molecule.

[PROBLEM: Convince yourselves that using (11) instead of using (2) for the entropy does not change any aspect of our discussion of the thermodynamics of an ideal gas, although it does necessitate a redefinition of the Helmholtz free energy: $\bar{F} = -kT\ln(\frac{z^N}{N!})$.]

[PROBLEM: Calculate a numerical value of the entropy of Helium at room temperature using eq.(11) instead of eq. (2): compare the result with your earlier calculation using eq. (2).]

Let us now use our new definition of the entropy of a gas to study the mixing of two volumes of gas. Then the significant bits of z are given by

$$z \propto (mkT)^{3/2}V \tag{12}$$

where m is the molecular mass.

If we consider mixing two volumes of the same gas at the same temperature, then initially

$$\begin{aligned} \bar{S}_1 &= N_1k\{\ln\frac{V_1}{N_1} + \frac{3}{2}\ln mkT + \text{constants}\} \\ \bar{S}_2 &= N_2k\{\ln\frac{V_2}{N_2} + \frac{3}{2}\ln mkT + \text{constants}\} \end{aligned} \tag{13}$$

and from (11) again

$$\bar{S}_{12} = (N_1 + N_2)k\{\ln(\frac{V_1 + V_2}{N_1 + N_2}) + \frac{3}{2}\ln mkT + \text{constants}\}$$

and this is NOT the sum of the entropies of N_1 in $V_1 + V_2$ and N_2 in $V_1 + V_2$!

If $\frac{N_1}{V_1} = \frac{N_2}{V_2}$ (e.g. $N_1 = N_2 = N$, $V_1 = V_2 = V$) we have $\bar{S}_{12} = S_1 + S_2$. If after the mixture has reached equilibrium we slide in a partition, V_1 contains N_1' and V_2 contains V_2', such that $N_1'/V_1 = N_2'/V_2$ and eq. (11) indeed gives no change of entropy.

Suppose we mix two volumes of different gases (again taking them at the same temperature for simplicity)

$$\bar{S}_1 = N_1k\{\ln\frac{V_1}{N_1} + \frac{3}{2}\ln m_1kT \;.....\}$$

$$\bar{S}_2 = N_2k\{\ln\frac{V_2}{N_2} + \frac{3}{2}\ln m_2kT \;.....\}$$

In the new volume these two different gases exert their own partial pressures,

[PROBLEM: Prove it.]

and

$$\begin{aligned}\bar{S}_{12} &= N_1k\{\ell n\frac{V_1 + V_2}{N_1} + \frac{3}{2}\ell n m_1kT \ldots\ldots\} \\ &+ N_2k\{\ell n\frac{V_1 + V_2}{N_2} + \frac{3}{2}\ell n m_2kT \ldots\ldots\}\end{aligned} \tag{14}$$

when $\bar{S}_{12} - (S_1 + S_2) = N_1k\ell n(\frac{V_1 + V_2}{V_1}) + N_2k\ell n(\frac{V_1 + V_2}{V_2})$

$$\rightarrow (N_1 + N_2)k\ell n2 \quad \text{if} \quad V_1 = V_2$$

Let the thing settle down and slide in a partition. The partial pressures are now uniform and we find in V_1 N_{11} of molecule 1, N_{12} of molecule 2, and in V_2 N_{21} of molecule 1, N_{22} of molecule 2, such that $N_{11}/V_1 = N_{21}/V_2$ and $N_{12}/V_1 = N_{22}/V_2$. It is at once clear that the entropy is still given by (14).

So our new definition of entropy preserves the ideal gas thermodynamics, and solves all problems associated with the entropy of mixing gases. Furthermore, the Sakur-Tetrode equation is right. The only difficulty is that we had to inject in a semi-classical way the idea of the indistinguishability of molecules, and classically this is almost incomprehensible. (Classically it is of course always open to us to argue that we don't know the constants in the expression for entropy and simply subtract $N\ell nN - N$ in an ad hoc fashion.) If you think of a swarm of classical identical particles, you can pick them out of a box one by one and put them into a new box, thereafter following their trajectories. The identical particles may be distinguished by the trajectories they follow, so classically our original scheme of counting microstates was surely correct? But remember that our molecules obey quantum mechanics (we had an example of this in the freezing out of degrees of freedom) and each has to be represented by a wave function. In a collision the wave packets coalesce and separate again, and if the particles are identical, who can say which is which after they emerge from the collision? We must search for the origin of the indistinguishability of identical particles in quantum mechanics.

The argument suggests that our original recipe was correct for localised particles. If we bring together two crystals, the atoms are locked in the lattices and cannot move about. There is no entropy of mixing and our original formulation didn't give one (the volume of the crystal didn't affect the energy levels). If the atoms are locked on lattice sites, there are no close collisions in which the wave functions become entangled and so indeed we would expect such a situation to correspond to distinguishable particles, and our original method of counting to be correct. Thus we expect eq. (2) to hold for LOCALISED particles and eq. (11) to hold for UNLOCALISED particles.

[PROBLEM: Use eq. (2) to find the entropy in the Einstein model of a solid (Lecture 6). Show that the low temperature limit is

$$S = 3Nk\frac{\hbar\omega}{kT}e^{-\hbar\omega/kT} \xrightarrow[T \to 0]{} 0$$

Does this result make you worry about our expression eq. (11) for the entropy of a gas?]

Lecture 10

INDISTINGUISHABILITY AND EXCHANGE DEGENERACY

We have an operational definition already of what we must mean by indistinguishability: if a set of identical particles are indistinguishable then every microstate is a macrostate (excluding degeneracy or the clumping of levels for convenience). Classically this doesn't make sense but the overlap of wave functions in quantum mechanics makes it at least seem possible, so we seek for a reason in quantum mechanics.

The energy levels of a system of identical particles cannot depend on which particle is where, whether or not they are distinguishable. The Hamiltonian describing the system must be symmetric under any permutation of the particle coordinates. We have been considering a system with a lot of weakly interacting components, the interactions not affecting the component energy levels but merely allowing thermal equilibrium to be achieved. The Hamiltonian is

$$H = \Sigma H_i \qquad H_i = T_i + V_i = T(r_i) + V(r_i) \tag{1}$$

If $\psi^k(r_i)$ is the wavefunction of the kth state for particle i, then

$$H_i\psi^k(r_i) = E_k\psi^k(r_i)$$

and the function

$$\psi^k(r_i)\psi^\ell(r_j)\psi^m(r_s)\ldots \tag{2}$$

is a solution of (1) with $E = E_k + E_\ell + E_m \ldots\ldots$ (3)

With the Hamiltonian symmetric under any permutation of the particles, any permutation of (2) will be a solution with the eigenvalue (3) and so will any linear combination of the permuted solutions. All these states are degenerate and if the particles are truly identical this degeneracy is absolute - it isn't like accidental degeneracies in, for example, the hydrogen atom which are always removed if you probe deeply enough.

If the particles are indistinguishable all these permutations must be projections of the same physical state. Our requirement that two of the particles are indistinguishable can be expressed by demanding that the expectation value

$$\langle\psi|R|\psi\rangle$$

of any operator R is to be unchanged if we swop any pair of particles and this is ensured if

$$\psi(i,j) = \pm\psi(j,i)$$

For two particles the functions

$$\psi^k(r_i)\psi^\ell(r_j) \pm \psi^k(r_j)\psi^\ell(r_i)$$

have this property - but an arbitrary linear combination of the two does NOT. These two states are eigenfunctions of the permutation operation P_{ij}; since H is symmetric for identical particles and

$$H\psi = i\hbar\frac{\partial\psi}{\partial t}$$

if we start in a symmetric state we have a symmetric state for ever after and if we start with an antisymmetric state the state remains antisymmetric as it evolves with time. We may not have a mixture of symmetric and antisymmetric states if our indistinguishability condition is to be met. Thus if two particles are indistinguishable, the wave function must be either symmetric or antisymmetric under their interchange.

In non-relativistic quantum mechanics we remove the exchange degeneracy and allow for indistinguishability by adding a POSTULATE: the wave function for a set of identical particles is either totally symmetric OR totally antisymmetric under any permutation.
[NOTE: for a two particle state the symmetric and antisymmetric combinations exhaust the possible classes of permutation symmetry. For three or more particles there will be further distinct combinations, symmetric under interchange of some pairs and antisymmetric under interchange of others. (The number of permutations is N!) These

possibilities do not seem to be realised in nature, and it is appealing to suppose that if a wave function is symmetric in any one pair of identical particles, then it is symmetric under interchange of any other pair of the same particles.]

There exists just one totally symmetric state

$$\psi_S = \Sigma P \psi^k(r_i)\psi^\ell(r_j)\ldots\ldots \quad (4)$$

and just one totally antisymmetric state, which can be written

$$\psi_A = \begin{vmatrix} \psi^k(r_i) & \psi^k(r_j) & \psi^k(r_s)\ldots\ldots \\ \psi^\ell(r_i) & \psi^\ell(r_j) & \psi^\ell(r_s)\ldots\ldots \\ \psi^m(r_i) & \psi^m(r_j) & \psi^m(r_s)\ldots\ldots \end{vmatrix} \quad (5)$$

Note that if any pair of the state indices k, ℓ, m ... in (5) have the same value, then (5) vanishes. This is a mathematical expression of the Pauli exclusion principle. In the end of course we have to appeal to experiment to tell us whether or not particles are really identical - perhaps electrons come each stamped with the maker's number, even though we don't know where to find it. Experimentally particles of half integral spin $\frac{1}{2}\hbar$, $\frac{3}{2}\hbar$...) have totally antisymmetric wave functions, and obey the Pauli principle, and particles of integral spin (0, $\hbar$, $2\hbar$....) have totally symmetric wave functions. There is no explanation for this in non-relativistic quantum mechanics, although an explanation does exist in quantum field theory.

Localised Identical Particles

Suppose we now consider identical particles on a lattice, electrons in atoms bound in a lattice or the atoms themselves. We might write

$$H_i = T_i + V(r_i - r_a) \qquad r_a, r_b \text{ specify}$$

$$H_j = T_j + V(r_j - r_b) \qquad \text{lattice sites}$$

These two pieces of the Hamiltonian are NOT symmetric in r_i and r_j and indeed the solutions for $H = H_i + H_j$ are respectively $\psi_a^k(r_i)\psi_b^\ell(r_j)$ and $\psi_a^\ell(r_i)\psi_b^k(r_j)$ which are distinguishable degenerate solutions. How do we reconcile this perfectly sensible result with the postulate that wave functions for identical particles are either symmetric or antisymmetric?

We should recognise that particle i could be assigned to any site, and write

$$H_i = T_i + V(r_i - r_a) + V(r_i - r_b)\ldots$$

and then H will be totally symmetric. Possible solutions for two particles will then be (provided the potential localises particles)

$$\psi_a^k(r_i)\psi_a^\ell(r_j),\ \psi_a^\ell(r_i)\psi_a^k(r_j),\ \psi_b^k(r_i)\psi_b^\ell(r_j),\ \psi_b^\ell(r_i)\psi_b^k(r_j)$$

$$\psi_a^k(r_i)\psi_b^\ell(r_j),\ \psi_a^k(r_j)\psi_b^\ell(r_i),\ \psi_a^\ell(r_i)\psi_b^k(r_j),\ \psi_a^\ell(r_j)\psi_b^k(r_i)$$

From the states in the second line we can make TWO pairs of symmetric states or two pairs of antisymmetric states in general. They are distinguishable - the thing at site a is in state k or the thing at site b is in state k - provided the potentials localise the wavefunctions.

The two pairs are

$$\psi_a^k(r_i)\psi_b^\ell(r_j) \pm \psi_a^k(r_j)\psi_b^\ell(r_i) \quad \text{and}$$

$$\psi_a^\ell(r_i)\psi_b^k(r_j) \pm \psi_a^\ell(r_j)\psi_b^k(r_i)$$

The four solutions on the top line also split into two pairs, but correspond to two particles on the same lattice site - all right for electrons bound in atoms but not for a description of the atoms themselves where the interatomic forces become strongly repulsive at short range (because of the Pauli principle). In either case, with two particles and two sites we obtain two distinguishable symmetric solutions or two distinguishable antisymmetric solutions because the sites are distinguishable even though the particles are not. We may therefore treat localised particles as if they are distinguishable, and our enumeration of available states for localised particles, made under the assumption of distinguishability, remains correct.

[PROBLEM: Why do we have to treat molecules in a gas as identical and indistinguishable, while when tossing two pennies we expect (and find) HH in $\frac{1}{4}$ of the cases, TT in $\frac{1}{4}$ of the cases, and an H and a T in $\frac{1}{2}$ of the cases?]

Unlocalised Identical Particles - gases

We now have enough information to work out properly the number of states accessible to a gas of identical particles, subject to the proviso that the occupation numbers of individual states are small.

There is only one microstate for every macrostate, because of the indistinguishability of the particles, and we must therefore resort to the trick of grouping the energy levels in clumps of g_i states. Each clump has an occupation number m_i, which though large is much smaller than g_i.

If our particles have symmetric wave functions, there is no restriction on the number of particles in any one level. We must calculate the number of ways of placing m_i particles in g_i states, with no restriction on the occupation numbers and no distinction among the particles.

Imagine a line of m_i particles, and a set of g_i-1 barriers which may be lowered to divide them into g_i groups. How many ways are there of arranging these $m_i + g_i - 1$ objects on the line? The answer is $(m_i + g_i - 1)!$ ways. BUT this includes $m_i!$ permutations of the identical particles (without physical significance) and $(g_i - 1)!$ permutations of the $g_i - 1$ barriers (also without physical significance). So the total number of distinguishable arrangements is

$$\frac{(m_i + g_i - 1)!}{m_i!(g_i - 1)!}$$

If instead the wave function is to be antisymmetric, we may only have ONE or ZERO particles in a given state. The number of physically significant states is then given by the number of ways of selecting m_i (filled) states from g_i available. This is simply

$$g_i(g_i - 1) \;.....\; (g_i - m_i + 1)$$

$$= \frac{g_i!}{(g_i - m_i)!}$$

and this must be divided by the number of permutations of m_i particles to yield

$$\frac{g_i!}{m_i!(g_i - m_i)!}$$

Particles with integral spin are said to obey Bose-Einstein statistics, and are known as bosons. Particles with half integral spin are said to obey Fermi-Dirac statistics and are called fermions. The weights for gases of these two classes of particle are

$$W_{BE} = \Pi \frac{(m_i + g_i - 1)!}{m_i!(g_i - 1)} \tag{6}$$

$$W_{FD} = \Pi \frac{g_i!}{m_i!(g_i - m_i)!} \tag{7}$$

Note that if $g_i = 1$ then both W_{BE} and W_{FD} become unity: the clumping of levels is essential to extract anything useful by this method. (Note also that the total number of particles does not appear in (6) or (7).) Maximise W_{BE} with respect to m_i, subject to $\Sigma m_i = N$, $\Sigma \varepsilon_i m_i = E$

$$\ell n W_{BE} = \Sigma\{(m_i + g_i - 1)\ell n(m_i + g_i - 1) - m_i \ell n m_i - (g_i - 1)\ell n(g_i - 1)\}$$

$$\frac{d\ell n W_{BE}}{dm_i} + \alpha + \beta\varepsilon_i = 0$$

yields $$\frac{g_i - 1}{m_i} = e^{-(\alpha+\beta\varepsilon_i)} - 1$$

or $$m_i \simeq \frac{g_i}{e^{-(\alpha+\beta\varepsilon_i)} - 1}$$ Bose-Einstein distribution

while maximising W_{FD} yields

$$\frac{g_i - m_i}{m_i} = e^{-(\alpha+\beta\varepsilon_i)}$$

$$m_i \simeq \frac{g_i}{e^{-(\alpha+\beta\varepsilon_i)}+1}$$ Fermi-Dirac distribution

so that $$\bar{n}_i(BE) = \frac{1}{e^{-(\alpha+\beta\varepsilon_i)}-1}$$

$$\bar{n}_i(FD) = \frac{1}{e^{-(\alpha+\beta\varepsilon_i)}+1}$$

and if these averages are small they reduce to the classical Boltzmann distribution.

[PROBLEM: Investigate the entropy of BE and FD gases in this classical regime.]

These are the distributions we obtained in the second lecture by considering states of equilibrium. Note that the results are still well defined in the limit $g_i \to 1$, but the derivation is only valid for large g_i, m_i and in the case of Fermi-Dirac statistics, large $g_i - m_i$. We can't trust the derivation (even though the results are correct) except in the classical regime. It is just when the level spacing gets to be important that we expect quantum effects to arise and so just where the results get exciting (degenerate Fermi gases and the Bose-Einstein condensation) the treatment breaks down! We would like a better way of proceeding, and fortunately it exists.

Lecture 11

THE BETTER WAY: CANONICAL AND GRAND CANONICAL APPROACHES

We have so far considered an isolated system with specified energy and a specified number of particles (or other components). Such a system is known as a microcanonical ensemble. The weaknesses of such a development of statistical mechanics are:-

1. specified energy
2. specified number of particles
3. the need to assume only weak interactions among the components.

We would like a treatment valid when none of these restrictions hold and we would also like a treatment which overcomes the inadequacy of Stirling's theorem when considering Bose-Einstein and Fermi-Dirac gases.

The better way comes in two versions - the canonical treatment and the grand canonical treatment. In the former case we consider a system in thermal contact with a huge reservoir - now the energy of the system isn't fixed. In the latter case the system considered is in both thermal and diffusive contact with a huge reservoir, and now neither the energy of the system nor the number of components is fixed. Contact with the reservoir is made weakly, for example through the walls or surfaces bounding a macroscopic volume, so that only a small number of components are disturbed. The system thus retains its identity and contact with the reservoir does not distort the energy levels of the system. The components of the system may however be allowed to interact strongly if desired. The reason is that we only need to specify the energy levels of the whole system (NOT the component energy levels) regardless of how we are supposed to compute them. We are going to treat our whole macroscopic system as a single component of a larger system. [Remember that we were able to calculate mean occupation numbers for the energy levels of a single component and the mean component energy for components in the microcanonical approach.]

The canonical approach is more appropriate than the microcanonical approach for a system in contact with a heat reservoir, and will give us at least FORMAL results which apply even when components interact strongly. It has some computational advantages. The grand canonical approach is appropriate when the number of components really isn't fixed - for example in chemistry, or in the cores of stars, or when we are looking at only a little bit of a system, as we often do. When the number of components is fixed the method still has computational advantages and may be used provided the distribution of number of components is sharp. (See for example Rushbrooke - Statistical Mechanics - Ch. XVII.)

Start by being very simple minded. We have a system with a fixed number of components, in contact with a heat reservoir. We want the probability that it is in a state of energy E_i (E_i is the system energy, NOT a component energy). One way of doing it (particularly appropriate if we are looking at one bit of a large system) is to consider the reservoir to be made from a large number of copies of our system, all weakly interacting, and with a total energy $\mathcal{E}$. Each of these systems is identical and localised and we suppose we know their SYSTEM energy levels: we may now apply the microcanonical approach, Stirling's theorem and all, without any difficulty. If there are N_i systems with energy E_i, then the number of microstates making up this macrostate is

$$\mathcal{W} = N!\Pi \frac{g_i^{N_i}}{N_i!}$$

maximise $\ell n\mathcal{W}$ subject to the constraints $\Sigma N_i = N$ and $\Sigma N_i E_i = \mathcal{E}$ and obtain

$$N_i = \frac{N g_i e^{-E_i/kT}}{\Sigma e^{-E_i/kT}}$$

and the probability that our single system is to be found in a single state of energy E_i is given by

$$P_i = \frac{e^{-E_i/kT}}{\Sigma e^{-E_i/kT}} \qquad (1)$$

The function $$Z = \Sigma e^{-E_i/kT} \qquad (2)$$

is summed over all system energy levels and is the system partition function, not to be confused with the component partition function which I have always designated by z.

The mean energy of our system, which we expect to be rather well defined, is

$$\bar{E} = \Sigma P_i E_i \qquad (3)$$

(This is NOT the huge total energy of our ensemble of systems.)

Now
$$\begin{aligned}
\ell n W_{max} &= N\ell nN - N + \Sigma N_i \ell n g_i - \Sigma N_i \ell n N_i + \Sigma N_i \\
&= N\ell nN - \Sigma N_i \ell n \left(\frac{N_i}{g_i}\right) \\
&= N\ell nN - \Sigma \frac{N g_i e^{-E_i/kT}}{Z} \ell n \frac{N e^{-E_i/kT}}{Z} \\
&= -N\Sigma \frac{g_i e^{-E_i/kT}}{Z} \ell n \frac{e^{-E_i/kT}}{Z} \\
&= -N\Sigma P_i \ell n P_i
\end{aligned}$$

So the contribution of just one of the N systems (our system) to $\ell n W_{max}$ is

$$\ell n W = -\Sigma P_i \ell n P_i \qquad (4)$$

(where degenerate states must be explicitly included in the sum).

We have of course

$$d\bar{E} = \Sigma P_i dE_i + \Sigma E_i dP_i$$

for our single system. The first term is WORK, the second is HEAT.

Now the system energy levels E_i change in response to some external constraint X (Volume, magnetic field ...). Differentiate the SYSTEM partition function Z with respect to X at constant T:

$$Z = \Sigma e^{-E_i/kT}$$

$$\left(\frac{\partial \ell nZ}{\partial x}\right)_T = \frac{1}{Z}\left(\frac{\partial Z}{\partial X}\right)_T = -\frac{1}{kT}\Sigma\frac{\partial E_i}{\partial X}\frac{e^{-E_i/kT}}{Z}$$

$$= -\frac{1}{kT}\Sigma P_i\frac{\partial E_i}{\partial X}$$

The work term can be represented as

$$đW = +\left(\frac{\partial F}{\partial X}\right)_T dX$$

where $F = -kT\ell nZ$ - we have a Helmholtz free energy for our system.

In this whole setup the conditions for $\ell n W_{max} \approx \Sigma\ell n W$ are satisfied, and so we expect that the contribution of our one system to the entropy will be

$$S = -k\Sigma P_i \ell n P_i$$ from (4) (compare with results listed at the end of Lecture 7).

It is then trivial to show that

$$\bar{E} = F + TS$$

$$d\bar{E} = \left(\frac{\partial F}{\partial X}\right)_T dX + TdS = đW + đQ$$

[PROBLEM: It isn't so trivial as to be not worth showing, so show it! Refer to Lecture 7 if necessary.]

We now have the laws of thermodynamics without any sort of restriction to systems of weakly interacting components. $\bar{E}$, S, F can all be calculated from the system partition function Z and to calculate Z we need a knowledge of the energy levels of the system. Usually we can only calculate Z for systems of weakly interacting components, or by making approximations, but even if we cannot calculate Z we have justified the application of thermodynamics to arbitrary systems. (We do however need to know that $\bar{E}$ is sharply defined).

We may now make a guess at what will happen if the number of components in our system is variable. It is to be presumed that the energy (and the energy levels) of our system will change if the number of components changes, so we expect that the probability of finding the system in a state with n components and energy $E_i(n)$ will be given by

$$P_i \propto e^{[\mu n - E_i(n)]/kT}$$

where μ is some kind of work function (it is called the chemical potential). The normalisation factor for P_i would then be

$$\mathcal{Z} = \sum_{n,i} e^{[\mu n - E_i(n)]/kT} \tag{5}$$

which is known as the grand partition function, or grand sum.

So with this introduction, let's do it all properly. Consider a reservoir, this time of unspecified construction except that it is to be large in comparison with our system. Our system is in thermal contact with it and may be in diffusive contact so that components can be exchanged.

The reservoir contains N components and has energy E, our system has n components and energy ε; $N \gg n$, $E \gg \varepsilon$. $N + n = N_o$, $E + \varepsilon = E_o$. Note at this point that by $E + \varepsilon$ = constant we mean that the total energy of system + reservoir is well defined down to the limits imposed by irreducible interactions with the outside world, and by the uncertainty principle. An excursion of ε well beyond these limits must be balanced by a corresponding excursion of E.

Let the number of microstates corresponding to a given macrostate of the <u>system</u> be $w(n,\varepsilon)$ and the number corresponding to a given macrostate of the <u>reservoir</u> be $W(N,E)$. The number of corresponding states of the system + reservoir is then given by the product

$$\mathcal{W}(N,n,E,\varepsilon) = w(n,\varepsilon)W(N,E)$$

and the total number of microstates of system + reservoir is

$$\mathcal{W} = \Sigma w(n,\varepsilon)W(N,E)$$

where the sum is to be taken over all values of n,N,ε,E consistent with $n + N = N_o$, $\varepsilon + E = E_o$.

Find the maximum term in this sum:

$$d(wW) = wdW + Wdw$$

$$\doteq w\{\frac{\partial W}{\partial N}dN + \frac{\partial W}{\partial E}dE\} + W\{\frac{\partial w}{\partial n}dn + \frac{\partial w}{\partial \varepsilon}d\varepsilon\}$$

Now $dN + dn = 0$ and $dE + d\varepsilon = 0$ so

$$d(wW) = (w\frac{\partial W}{\partial N} - W\frac{\partial w}{\partial n})dN + (w\frac{\partial W}{\partial E} - W\frac{\partial w}{\partial \varepsilon})dE$$

and wW has an extreme value when

$$w\frac{\partial W}{\partial N} = W\frac{\partial w}{\partial n} \quad ; \quad w\frac{\partial W}{\partial E} = W\frac{\partial w}{\partial \varepsilon} \tag{6}$$

Divide (6) by wW and obtain

$$\frac{\partial \ln w}{\partial n}\bigg|_e = \frac{\partial \ln W}{\partial N}\bigg|_e \quad ; \quad \frac{\partial \ln w}{\partial \varepsilon}\bigg|_e = \frac{\partial \ln W}{\partial E}\bigg|_e \tag{7}$$

Using the statistical assumption that the state of system + reservoir which has the greatest number of microstates is the most probable, we use the equations (7) as the definition of equilibrium. But remember that equilibrium is an average state ..... see last PROBLEM for this lecture.

Now suppose that the system is in some specified state with energy ε_i, while in contact with the reservoir. The number of states of system + reservoir is just the number of states of the reservoir compatible with the chosen state of the system and the constraints (total number of components and total energy constant).

Then $P(n_i, \varepsilon_i) \propto W(N_o - n_i, E_o - \varepsilon_i)$

using our statistical assumption again. The relative probability of two states of the system is given by

$$\frac{P(n_1, \varepsilon_1)}{P(n_2, \varepsilon_2)} = \frac{W(N_o - n_1, E_o - \varepsilon_1)}{W(N_o - n_2, E_o - \varepsilon_2)} = e^{S(1)-S(2)} \tag{8}$$

where $S(i) = \ln W(N_o - n_i, E_o - \varepsilon_i)$.

Now since $N >> n$ and $E >> \varepsilon$, N and E only change by very small fractional amounts for relatively large changes in n, ε. Therefore expand about the equilibrium value of N and E, N_e and E_e

$$S(i) = \ell nW(N_o - n_i, E_o - \varepsilon_i)$$
$$= \ell nW(N_e, E_e) + (N_o - N_e - n_i)\frac{\partial \ell nW}{\partial N}\bigg|_e + (E_o - E_e - \varepsilon_i)\frac{\partial \ell nW}{\partial E}\bigg|_e \tag{9}$$

when $$\frac{P(n_1,\varepsilon_1)}{P(n_2,\varepsilon_2)} = \frac{e^{n_1[-\frac{\partial \ell nw}{\partial n}|_e] - \varepsilon_1\frac{\partial \ell nw}{\partial \varepsilon}|_e}}{e^{n_2[-\frac{\partial \ell nw}{\partial n}|_e] - \varepsilon_2\frac{\partial \ell nw}{\partial \varepsilon}|_e}} \tag{10}$$

where we have used eqs. (7) to write (10) in terms of the properties of the system.

We now identify $\frac{\partial \ell nW}{\partial E}\big|_e = \frac{\partial \ell nw}{\partial \varepsilon}\big|_e$ with $^1/kT$ (11)

and we define

$$-\frac{\partial \ell nW}{\partial N}\bigg|_e = -\frac{\partial \ell nw}{\partial n}\bigg|_e = \frac{\mu}{kT} \tag{12}$$

so that μ has the dimensions of an energy, and is the chemical potential. (The quantity $e^{\mu/kT}$ is called the absolute activity). Note that both T and μ are set by the reservoir in contact with our system.

Before we go any further, note that the whole argument depends crucially on the reservoir NOT being in a state of perfectly well defined energy. We need W(N,E) to be a large number for a well defined system energy and unless you believe in huge absolute degeneracies this isn't possible unless the energy of the reservoir (or of system + reservoir) is spread over a large number of states about $E_o - \varepsilon$ (or E_o). For a macroscopic system the uncertainty principle assures this. (See the last problem at the end of Lecture 5.)

[PROBLEM: Refer to the last problem at the end of Lecture 5. How many energy states of the whole gas is the energy spread over 10^5 secs. after boxing it?]

Forget the reservoir and set $n_i \to N$, $\varepsilon_i \to E_i$. The relative probabilities in (10) are normalised by

$$\mathcal{Z} = \sum_{N,i} e^{N\mu/kT - E_i(N)/kT} \tag{13}$$

which is the grand partition function, or the grand sum.

Now let us identify the entropy as the number of states accessible to our system. We have the probability of finding our system in a state with N components and energy $E_i(N)$. How many states are accessible? The easiest thing to do is to use a frequency interpretation of probability and imagine M copies of our system. Then the number in a state with probability P_i is $m_i = MP_i$. The number of ways of realising a given set of m_i is just

$$W_{m_i} = M!\Pi\frac{1}{m_i!}$$

and the logarithm is

$$\ell nW_{m_i} = M\ell nM - M - \Sigma(m_i\ell nm_i - m_i)$$

$$= M\ell nM - M - \Sigma(MP_i\ell nMP_i - MP_i)$$

$\ell nW_{m_i} = -\Sigma MP_i\ell nP_i$ where the sum is such that $\Sigma P_i = 1$ and this is the logarithm number of states accessible to M systems distributed according to P_i.

Thus, since each of our M systems is equivalent, we identify

$$\ell nW_p = -\sum_{Ni}\Sigma P(N,E_i(N)\ell nP(N,E_i(N) \tag{14}$$

where $P(N,E_i(N)) = e^{N\mu/kT - E_i(N)/kT}/\mathcal{Z}$ (15)

[PROBLEM: Check that the number of accessible states for a die is given by $\ell nW_p = -\Sigma P\ell nP$.]

The entropy of our system is now

$$S = k\ell nw_p$$

The mean energy is

$$\bar{E} = \Sigma\Sigma E_i(N)P(N,E_i(N))$$

and the mean number of components is

$$\bar{N} = \Sigma\Sigma NP(N,E_i(N))$$

we may then obtain for reversible processes the relation

$$d\bar{E} = TdS + \mu d\bar{N} + dW \tag{16}$$

where the term $\mu d\bar{N}$ is new and represents chemical work.

[PROBLEM: This one is important

Obtain (16). Start by showing that

$$\bar{E} = \bar{N}\mu + TS - kT\ln\mathcal{Z}$$

The quantity $\Omega = -kT\ln\mathcal{Z}$ is called the grand potential and is the generalisation of the Helmholtz free energy. Remember that μ controls $\bar{N}$, as T controls $\bar{E}$ so take Ω as a function of μ, T, X where X is an external constraint affecting the energy levels. Show that

$$đW = \left(\frac{\partial\Omega}{\partial X}\right)_{\mu,T} dX$$

and hence obtain (16).

Now work back and show that if we identify $\ln w$ with $-\Sigma\Sigma P_i \ln P_i$, then (11) and (12) must read

$$\frac{\partial \ln w}{\partial \bar{E}} = \frac{1}{kT} \quad , \quad \frac{\partial \ln w}{\partial \bar{N}} = -\frac{\mu}{kT}$$

You shouldn't get stuck but consult Mandl, Ch. 12 or Kittel, Ch. 7, if you do.]

We have now established thermodynamics for a general system characterised by a temperature (the temperature of the reservoir) without any of the restrictions listed at the beginning of the lecture. The probabilities given by eq. (15) are valid even for a microscopic system, provided it is in contact with a macroscopic system! There remain a few points to be cleared up:

1. Why did we expand $\ln W$ and $\ln w$ in obtaining (8) instead of W and w?
2. w was a factor of only the maximum term in $\mathcal{W}$ and so does not represent all the states accessible to our system: how can we take $S = -k\ln w$? This will be all right if the most probable state is overwhelmingly the most probable, but
3. How can this be the case when probabilities are exponentials?

Lecture 12

MEAN VALUES AND FLUCTUATIONS

We calculated the probability of a system in thermal and diffusive contact with a reservoir being in a state with n_1 components and energy ε_1 by setting

$$P(n_1,\varepsilon_1) \propto W(N_o - n_1,\ E_o - \varepsilon_1) \tag{1}$$

where W is the number of states accessible to the reservoir, compatible with n_1, ε_1 for the system. We then expanded (1) about the equilibrium values, but made the expansion in ℓnW rather than in W itself. Why? Was it just to get exponential probabilities? (NO!) There are two answers. The first is that our equilibrium conditions were already expressed in terms of derivatives of ℓnW and ℓnw, so it was a natural thing to do. The second answer is that an expansion of W is insufficiently accurate. We may see this by making a rough estimate of the number of states accessible to a macroscopic system. Let the system have N components and the energy levels of the components are quantised. Suppose that these levels have equal spacing for simplicity. The separation is Δ: let component i have energy $n_i\Delta$. The total energy is

$$E = \sum_i n_i\Delta$$

where i runs from 1 to N. (Note: n_i is NOT the number of components in a given energy level.)

The maximum value of any of the n_i's is E/Δ, and there are N values of i. The variation with E of the total number of energy levels of the whole system less than E will therefore be given approximately by the quantity $(\frac{E}{\Delta})^N$, representing a volume cut out of an N-dimensional lattice. This number, while giving the dependence of the number of states of energy $<E$ as a function of E for N components, is clearly an overestimate. Each component cannot simultaneously have energy E. We can improve the estimate by noting that the mean component energy is E/N and then take the volume of a lattice extending to $E/\Delta N$ along any axis. The number of states with energy $<E$ is then approximately

$$(\frac{E}{N\Delta})^N$$

and the number of states between E and E + δE is approximately

$$W \approx (\frac{E}{N\Delta})^{N-1} \frac{\delta E}{\Delta} \qquad (2)$$

The logarithm of this number of states is

$$\ell nW \approx (N-1)\ell n\frac{E}{N\Delta} + \ell n\frac{\delta E}{\Delta} \qquad (3)$$

The dependence of ℓnW on E is through a term $N\ell nE$ for a macroscopic state. If we expand this expression about, say, an equilibrium value of E,

$$E = E_e + E - E_e$$

and $$\ell nE \to \ell nE_e + \ell n(1 + \frac{E - E_e}{E_e}) \approx \ell nE_e + \frac{E - E_e}{E_e} \quad .....$$

Provided $E - E_e << E_e$, a first order expansion is sufficient. However, if we expanded W itself we would have to expand

$$E^N = (E_e + E - E_e)^N = E_e^N(1 + \frac{E-E_e}{E_e})^N$$

$$= E_e^N(1 + N(E - E_e)/E_e \dots)$$

and the condition that this expansion is valid is

$$N(E - E_e) << E_e$$

which is much tighter: $E_e/N \sim kT$ and a restriction $E - E_e << kT$ is intolerable. [For other discussions of this point see Mandl, p56, Kittel, p85.] The thermal reservoir is macroscopic regardless of the nature of the system so the derivation of probabilities is valid even for microscopic systems.

We are now in a position to tackle the problem of fluctuations in an approximate way. Let our system have energy E and consist of N components. The probability of the system having energy E, keeping N constant, is given by the Boltzmann factor

$$e^{-E/kT}$$

The number of states between E and E + δE is roughly proportional to

$$\left(\frac{E}{N\Delta}\right)^{N-1}\frac{\delta E}{\Delta} \tag{2}$$

The probability of finding the system between E and E + δE is then given by the number of states between E and E + δE, each multiplied by the probability of occupancy, so it is

$$P(E)\,\delta E \approx \left(\frac{E}{N\Delta}\right)^{N-1}\frac{\delta E}{\Delta}e^{-E/kT} \tag{4}$$

The term $\left(\frac{E}{\Delta}\right)^N$ grows hugely with E if N is a large number, while the exponentia term falls. $P(E)$ has a maximum which will be sharply defined. Write

$$P(E) \sim e^{+N\ell nE\ -\ E/kT}$$

The maximum occurs at

$$\frac{\partial}{\partial E}\{N\ell nE - {}^{E}/kT\} = 0 \qquad \text{i.e.} \qquad E = NkT!$$

This result reassures us but shouldn't surprise us. The interesting thing to do is to study the variation of $P(E)$ with small excursions of the energy

$$N\ell nE - {}^{E}/kT = N\ell n(E_{max} + \Delta E) - (E_{max} + \Delta E)/kT$$

Expand to second order:

$$= N\ell nE_{max} - E_{max}/kT + N\ell n\left(1 + \frac{\Delta E}{E_{max}}\right) - \Delta E/kT$$

$$= N\ell nE_{max} - E_{max}/kT + \frac{N\Delta E}{E_{max}} - {}^{\Delta E}/kT - \tfrac{1}{2}N\left(\frac{\Delta E}{E_{max}}\right)^2 \ldots\ldots$$

Since E_{max} = NkT the first order term vanishes, and

$$P(E_{max} + \Delta E) = P(E_{max})e^{-\frac{1}{2}N\left(\frac{\Delta E}{E_{max}}\right)^2} \tag{5}$$

so the width of this approximately gaussian distribution is $\sim \frac{E_{max}}{\sqrt{N}}$ and this is the size of the fluctuations in energy which occur. They are utterly negligible for a macroscopic body with $N \sim 10^{23}$. The energy is very precisely defined, despite the exponential variation of probability.

Fluctuations in the number of components may be studied through the same approximation. Write

$$P(N) \approx e^{\mu N/kT}\left(\frac{E}{N\Delta}\right)^{N-1} = e^{\mu N/kT + (N-1)\ell n\left(\frac{E}{N\Delta}\right)}$$

The maximum with respect to N occurs when

$$\frac{\partial}{\partial N}\left\{\frac{\mu N}{kT} + (N-1)[\ell nE - \ell nN - \ell n\Delta]\right\} = 0$$

Let the value of N for which this equation is satisfied be N_{max}, set $N = N_{max} + \Delta N$ and expand up to second order in ΔN. Terms in ΔN vanish because of the maximum, and we find

$$P(N) \approx P(N_{max})e^{-\frac{1}{2}\frac{(\Delta N)^2}{N_{max}}}$$

so that the fractional variation of N is expected to be negligible for a macroscopic object.

Provided that the fluctuations are small, the total number of states accessible to our system is indeed given with sufficient accuracy by the number of states accessible to the most probable macrostate: we discussed this in a different context in Lecture 4.

There are circumstances in which macroscopic systems exhibit large fluctuations (see Reidi, p151, Mandl, pp 59-60, 241) and we will never find these circumstances from the elementary considerations above. Now that we have illuminated the problem of fluctuations in a physical way, we develop a method for estimating fluctuations for real macroscopic systems by proceeding more formally.

Suppose that some quantity q has an approximately gaussian distribution

$$P(q) \sim e^{-\frac{(q-q_o)^2}{2\Delta^2}}$$

The average value of q is

$$\bar{q} = \frac{\int qP(q)\,dq}{\int P(q)\,dq} = \frac{\int q e^{-\frac{(q-q_o)^2}{2\Delta^2}} dq}{\int e^{-\frac{(q-q_o)^2}{2\Delta^2}} dq} = q_o$$

The average value of q^2 is (setting $\alpha = \frac{1}{2\Delta^2}$)

$$\frac{\int q^2 P(q)\,dq}{\int P(q)\,dq} = \frac{\int q^2 e^{-\alpha(q-q_o)^2} dq}{\int e^{-\alpha(q-q_o)^2} dq}$$

Let $q - q_o = x$ when

$$\overline{q^2} = \frac{\int (x + q_o)^2 e^{-\alpha x^2} dx}{\int e^{-\alpha x^2} dx} = \frac{\int x^2 e^{-\alpha x^2} dx}{\int e^{-\alpha x^2} dx} + q_o^2$$

The integrals are to be taken between $\pm\infty$ approximately and

$$\int x^2 e^{-\alpha x^2} dx = \frac{1}{2}\sqrt{\frac{\pi}{\alpha}}\frac{1}{\alpha}$$

$$\int e^{-\alpha x^2} dx = \sqrt{\frac{\pi}{\alpha}}$$

The ratio equals $\frac{1}{2\alpha} \rightarrow \Delta^2$ so that

$$\Delta^2 = \overline{q^2} - q_o^2 = \overline{q^2} - \bar{q}^2 \tag{6}$$

we thus estimate the mean square fluctuation of any quantity q by calculating the quantity

$$\Delta^2 = \overline{q^2} - \bar{q}^2$$

We may now apply this to an estimation of the fluctuations in a real physical system. If we want the fluctuation on particle number, then start from the definitions of the mean number $\bar{N}$ and the mean square number $\overline{N^2}$:

$$\bar{N} = \frac{\Sigma\Sigma N e^{\mu N/kT - E_i/kT}}{\mathcal{Z}} \tag{7}$$

$$\overline{N^2} = \frac{\Sigma\Sigma N^2 e^{\mu N/kT - E_i/kT}}{\mathcal{Z}} \tag{8}$$

Now remember that the grand sum $\mathcal{Z}$ is given by

$$\mathcal{Z} = \Sigma\Sigma e^{\mu N/kT - E_i/kT}$$

so we can write

$$\bar{N} = \frac{kT}{\mathcal{Z}}\frac{\partial \mathcal{Z}}{\partial \mu} \tag{9}$$

$$\overline{N^2} = \frac{k^2T^2}{\mathcal{Z}}\frac{\partial^2 \mathcal{Z}}{\partial \mu^2} \tag{10}$$

so $$\overline{N^2} - \bar{N}^2 = \frac{k^2T^2}{\mathcal{Z}}\left\{\frac{\partial^2 \mathcal{Z}}{\partial \mu^2} - \frac{1}{\mathcal{Z}}\left(\frac{\partial \mathcal{Z}}{\partial \mu}\right)^2\right\} \tag{11}$$

Now $$\frac{\partial \bar{N}}{\partial \mu} = \frac{1}{kT}\overline{N^2} - \frac{\bar{N}}{\mathcal{Z}}\frac{\partial \mathcal{Z}}{\partial \mu} \qquad \text{from (7)}$$

and $$\frac{\partial \mathcal{Z}}{\partial \mu} = \frac{\mathcal{Z}}{kT}\bar{N} \qquad \text{from (11)}$$

so (13) may also be written

$$\overline{N^2} - \bar{N}^2 = kT\frac{\partial \bar{N}}{\partial \mu} \tag{12}$$

and the fluctuations only become large if $\frac{\partial \bar{N}}{\partial \mu}$ is large, $\bar{N}$ enormously sensitive to small changes in μ. If we want the fluctuations in energy of our system,.we have

$$\bar{E} = \frac{\Sigma\Sigma E_i e^{\mu N/kT - E_i/kT}}{\mathcal{Z}} \tag{13}$$

$$\overline{E^2} = \frac{\Sigma\Sigma E_i^2 e^{\mu N/kT - E_i/kT}}{\mathcal{Z}} \tag{14}$$

Now for CONVENIENCE set $^1/kT = -\beta$, when holding μ/T constant

$$\bar{E} = \frac{1}{\mathcal{Z}}\frac{\partial \mathcal{Z}}{\partial \beta} \qquad \overline{E^2} = \frac{1}{\mathcal{Z}}\frac{\partial^2 \mathcal{Z}}{\partial \beta^2}$$

$$\overline{E^2} - \bar{E}^2 = \frac{1}{\mathcal{Z}}\left\{\frac{\partial^2 \mathcal{Z}}{\partial \beta} - \left(\frac{\partial \mathcal{Z}}{\partial \beta}\right)^2\right\}$$

$$\frac{\partial \bar{E}}{\partial \beta} = \overline{E^2} - \frac{\bar{E}}{\mathcal{Z}}\frac{\partial \mathcal{Z}}{\partial \beta}$$

and $$\frac{\partial \mathcal{Z}}{\partial \beta} = \mathcal{Z}\bar{E}$$

so that $$\overline{E^2} - \bar{E}^2 = \frac{\partial E}{\partial \beta} = \frac{\partial \bar{E}}{\partial T}\frac{\partial T}{\partial \beta}$$

$$T = -\frac{1}{k\beta} \quad \text{so} \quad \frac{\partial T}{\partial \beta} = \frac{1}{k\beta^2} = kT^2$$

and $$\overline{E^2} - \bar{E}^2 = kT^2\frac{\partial \bar{E}}{\partial T}$$ (and if $\bar{E} \sim NkT$ this is $\sim\bar{E}^2/N$)

and energy fluctuations only get large if $\bar{E}$ is enormously sensitive to T.

Since
$$\frac{\partial \bar{E}}{\partial T} = C_V$$
$$\overline{E^2} - \bar{E}^2 = kT^2 C_V$$

and this is another way of recognising when energy fluctuations become large: C_V very large. This sort of thing happens near critical points when we have phase transitions.

[See Reidi, Sect. 10.4 or a more advanced text such as Landau and Lifschitz - Statistical Physics, Huang-Statistical Mechanics.]

Lecture 13

IDENTICAL PARTICLES AGAIN

We may now use the grand canonical approach to obtain the properties of ideal gases of identical particles - Bose-Einstein and Fermi-Dirac statistics. The first thing we would like is the occupancy of any single particle state and this is now very easy to work out. Let the single particle energy level have single particle energy E and this single particle level is to be our system, in thermal contact with the rest of the gas. The system energy is thus NE, where N is the occupancy which is clearly NOT fixed.

Our general probability distribution for a system is given by

$$P = \frac{e^{\mu N/kT - E(N)/kT}}{\Sigma\Sigma e^{\mu N/kT - E(N)/kT}} \tag{1}$$

where the denominator is the grand partition function $\mathcal{Z}$. For our system consisting of just one single particle energy level with variable occupancy

$E(N) = NE$ and so

$$P = \frac{e^{N(\mu-E)/kT}}{\sum_N e^{N(\mu-E)/kT}}$$

and for this case

$$\mathcal{Z} = \sum_N e^{N(\mu-E)/kT} \tag{2}$$

Now for spin $\frac{1}{2}$ ... particles, fermions, the only possible occupation numbers are zero or one. We can write down Z at once

$$Z_{F-D} = 1 + e^{(\mu-E)/kT} \tag{3}$$

and since $\bar{N} = \Sigma NP$ we can write down the expression for $\bar{N}$ at once

$$\bar{N}_{F-D} = \frac{e^{(\mu-E)/kT}}{1+e^{(\mu-E)/kT}} \tag{4}$$

$$= \frac{1}{e^{(E-\mu)/kT}+1} \leqslant 1$$

and this is the distribution we obtained in the second lecture and found by dubious means via the microcanonical approach. REMEMBER that μ is the same for the system and for the reservoir in equilibrium, and may be regarded as being set by the reservoir. As the reservoir consists of all other single particle energy levels and any one of them may be chosen as the system, it is clear that μ is independent of which level we chose.

For integral spin particles, any number of identical particles are permitted in any level, so for Bose-Einstein statistics

$$Z_{B-E} = \sum_0^\infty e^{N(\mu-E)/kT} \tag{5}$$

This sum is easily evaluated, for

$$\sum_0^\infty x^n = \frac{1}{1-x}$$

(provided $x < 1$ and $e^{(\mu-E)/kT}$ must be <1 if Z_{B-E}, $\bar{N}$ are to converge)

$$Z_{B-E} = \frac{1}{1-e^{(\mu-E)/kT}}$$

$$\bar{N} = \sum_0^\infty \frac{Ne^{N(\mu-E)/kT}}{Z_{B-E}} \tag{6}$$

and $$\sum_0^\infty nx^n = x\frac{d}{dx}\sum_0^\infty x^n = x\frac{d}{dx}\frac{1}{(1-x)} = \frac{x}{(1-x)^2}$$

Thus
$$\bar{N} = \frac{e^{(\mu-E)/kT}}{1-e^{(\mu-E)/kT}}$$

$$\bar{N}_{B-E} = \frac{1}{e^{(E-\mu)/kT}-1} \qquad (7)$$

Note that μ must be common to any energy level. Eq. (7) is the Bose-Einstein distribution which we obtained in the second lecture and again without conviction using the microcanonical ensemble. The indistinguishability of the particles has been taken into account quite painlessly because our state was specified only saying that the occupancy of a given single particle level was N.

Our results (4) and (7) for the distribution of occupation numbers over the single particle levels in a Fermi-Dirac or Bose-Einstein gas are now quite generally valid so long as the gas is macroscopic: in particular the results hold even when the level spacing becomes comparable with or greater than kT. We may study not only the classical limit but also the limit where quantum effects are dominant.

[PROBLEM: Notice that (4) and (7) give the single particle level occupation numbers as a function of μ. You may therefore use the results from the last lecture to study the fluctuations in the occupancy of these levels - do so, and compare and contrast the F-D and B-E cases.]

We can now obtain the grand partition function for the whole gas. In one particular state of the gas there are n_i particles with energy E_i, so the whole gas has energy

$$E = \Sigma n_i E_i$$

$$N = \Sigma n_i$$

The sum over all possible states of the whole gas is made by summing over all possible values of each of the n_i's

$$\mathcal{Z} = \sum_{n_i} e^{\sum_i n_i(\mu-E_i)/kT}$$

$$= \prod_i \sum_{n_i} e^{n_i(\mu-E_i)/kT}$$

$$\mathcal{Z}_{F-D} = \prod_i \left(1+e^{(\mu-E_i)/kT}\right) \qquad (8)$$

for Fermi-Dirac statistics, or

$$\mathcal{Z}_{B-E} = \prod_i \frac{1}{(1-e^{(\mu-E_i)/kT})} \tag{9}$$

for Bose-Einstein statistics: the grand partition functions for the whole gas are just the products of the grand partition functions for the single particle levels.

Take the logarithm:

$$\ell n\mathcal{Z}_{F-D} = \sum_i \ell n(1 + e^{(\mu-E_i)/kT})$$

$$\ell n\mathcal{Z}_{B-E} = -\sum_i \ell n(1 - e^{(\mu-E_i)/kT})$$

and in the classical limit where the exponentials are small

$$\ell n\mathcal{Z} \simeq \sum_i e^{(\mu-E_i)/kT}$$

$$= e^{\mu/kT}\Sigma e^{-E_i/kT}$$

$$\to e^{\mu/kT}\int e^{-\varepsilon/kT}\,\frac{dn}{d\varepsilon}d\varepsilon$$

[PROBLEM: Why did I expand $\ell n\mathcal{Z}$ and not $\mathcal{Z}$?]

We may obtain everything we want from $\ell n\mathcal{Z}$. The integral, over single particle states, is one we have done before (Lecture 5) - call it x

$$x = \int e^{-\varepsilon/kT}\frac{dn}{d\varepsilon}d\varepsilon = \frac{4\pi\,(mkT)^{3/2}}{(2\pi\hbar)^3}\sqrt{\frac{\pi}{2}}\,V \quad \text{(times a spin factor)}$$

$$\ell n\mathcal{Z} = xe^{\mu/kT}$$

For example, defining

$$z_i = e^{\mu N/kT-E/kT} \qquad \mathcal{Z} = \Sigma z_i$$

(where E, N apply to the whole gas)

we have $\quad P_i = z_i/\mathcal{Z}$

and $\quad \bar{N} = kT\dfrac{\partial \ell n\mathcal{Z}}{\partial\mu}$

so $\quad \bar{N} = xe^{\mu/kT} \qquad$ and $\mu = kT\ell n\dfrac{\bar{N}}{x}$

which serves to determine μ in terms of $\bar{N}$.

Now remember that we found in the last lecture

$$\Delta N^2 = kT\frac{\partial\bar{N}}{\partial\mu}$$

$$\frac{\partial\bar{N}}{\partial\mu} = \frac{x}{kT}\,e^{\mu/kT} = \frac{\bar{N}}{kT} \qquad \text{so } \Delta N = \sqrt{\bar{N}}$$

and we have proved the expected results for number fluctuations in a gas!

[PROBLEM: Calculate the chemical potential for Helium gas at STP.]

The mean energy of a gas can be similarly obtained:

$$\bar{E} = \Sigma P_i E_i = \Sigma\frac{z_i E_i}{\mathcal{Z}} = kT^2\frac{\partial \ell n\mathcal{Z}}{\partial T} + \mu\bar{N}$$

$$\frac{\partial \ell n\mathcal{Z}}{\partial T} = e^{\mu/kT}\,\frac{3}{2}\frac{x}{T} - \frac{\mu}{kT^2}\,e^{\mu/kT}x$$

$$kT^2\frac{\partial \ell n\mathcal{Z}}{\partial T} = \frac{3}{2}\bar{N}kT - \mu\bar{N}$$

so $\qquad \bar{E} = \frac{3}{2}\bar{N}kT$

[PROBLEM: Calculate the energy fluctuations in this gas from $\mathcal{Z}$]

What about the entropy? This is defined as

$$S = -k\Sigma P_i \ell n P_i = -k\Sigma\frac{z_i}{\mathcal{Z}}\ell n\frac{z_i}{\mathcal{Z}}$$

$$= -k\{\frac{\mu\bar{N}}{kT} - \frac{\bar{E}}{kT} - \ell n\mathcal{Z}\}$$

$$= -\frac{\mu\bar{N}}{T} + \frac{\bar{E}}{T} + k\ell n\mathcal{Z} \quad (= \frac{\partial}{\partial T}\{kT\ell n\mathcal{Z}\})$$

$$= -k\bar{N}\ell n\frac{\bar{N}}{x} + \frac{3}{2}\bar{N}k + \bar{N}k$$

$$= \frac{5}{2}\bar{N}k - k\bar{N}\ell n(\frac{\bar{N}}{x})$$

which is the Sakur-Tetrode equation again (see Lecture 9, also Kittel, p167.)

We have obtained the correct expression for the entropy of an ideal gas, in the classical limit, from the grand canonical approach where we considered the energy levels of the whole system, because we counted single particle levels with N particles in them and that was taken as completely specifying the contribution to the grand sum.

All the properties of an ideal gas (with no internal degrees of freedom) have now been obtained quickly and cleanly from the grand partition function, obtained by summing

$$e^{\mu N/kT - E/kT}$$

over all the accessible states of the system. A real gas is complicated by the existence of intermolecular potentials which are a function of the separation of the molecules and affect the total energy of the system. Nevertheless, we can write down a partition function for such a gas and evaluate it by making successive approximations ... If you want to, you can read about this in Mandl, Section 7.8 or Rushbrooke, Chapter XVI.

Lecture 14

BOSONS: BOSE-EINSTEIN CONDENSATION AND BLACK BODY RADIATION

The crucial result we obtained for a gas of bosons is that the occupation number is distributed according to

$$\bar{n} = \frac{1}{e^{(\varepsilon-\mu)/kT}-1} \tag{1}$$

We have already studied the classical limit of such a gas. We are now interested in the opposite limit where quantum mechanics is important and the distribution function (1) departs from the value obtaining in the classical approximation. We can study this problem in an elementary way. (For a more advanced treatment, see Feynman - Statistical Mechanics, Ch. 1.)

In eq. (1) $\bar{n}$ is the average occupancy of a single particle level in a gas of non-interacting bosons, and ε is the energy of that level. The chemical potential μ is determined by the total number of particles in the gas.

Bose-Einstein Condensation

At absolute zero all the particles in the gas must be in the ground state: there is no exclusion principle for bosons. If N is the total number of particles, then

$$\lim_{T\to 0} \frac{1}{e^{(\varepsilon_o-\mu)/kT}-1} \to N \tag{2}$$

If N is to be large the exponential term must be close to unity, so expand it to first order

$$\lim_{T \to 0} \frac{1}{(\varepsilon_o-\mu)/kT} \to N$$

Then $\mu \to \varepsilon_o - \frac{kT}{N}$ as $T \to 0$ (3)

(in a great many treatments the ground state energy is taken as defining the zero of the energy scale. This is not necessary and I prefer to stick with the definition I have been using all along).

Suppose that $\mu \simeq \varepsilon_o - \frac{kT}{N}$ holds. Then in any other level $(\varepsilon-\mu)/kT = (\varepsilon-\varepsilon_o)/kT + \frac{1}{N}$

$$\bar{n} = \frac{1}{e^{(\varepsilon-\varepsilon_o)/kT + \frac{1}{N}} - 1} \tag{4}$$

If $\varepsilon-\varepsilon_o >> \frac{kT}{N} = \varepsilon_o-\mu$ then for all states other than the ground state we have a Bose-Einstein distribution with $\mu \simeq \varepsilon_o$, and when $^{\varepsilon}/kT >> 1$ we have a classical distribution. Consider for example an ideal gas of Helium, confined in a box of dimensions ℓ^3. In the ground state $\lambda = {}^h/p = 2\ell$ while in the first excited state $\lambda = {}^h/p = \ell$. Then

$$\varepsilon_1-\varepsilon_o = \frac{p_1^2 - p_o^2}{2M} = \frac{h^2}{2M\ell^2}\frac{3}{4} \simeq 10^{-18} \text{ eV}$$

if $\ell = 1$ cm.

Now $kT \simeq 10^{-4}T$ eV so that this condition is satisfied easily for a macroscopic value of $N \sim 10^{22}$.

If we go in steps of $\sim 10^{-18}$ eV, we certainly cannot use a Boltzmann distribution, but find

$$n \sim 10^{14}\, T \frac{4}{m^2-1} \tag{5}$$

where m is the number of half wavelengths in a 1 cm^3 box. The number of single particle states with momentum less than p is $\sim \frac{4\pi}{3}(\frac{p}{h})^3 V = \frac{\pi}{6}(\frac{m}{\ell})^3$ and the number of states around m varies as m^2 so we need to calculate carefully the number of particles in states other than the ground state.

$$N(\varepsilon > \varepsilon_0) \sim \int_0^\infty \frac{1}{e^{\varepsilon/kT}-1} \frac{dn}{d\varepsilon} d\varepsilon$$

$$= \frac{4\pi}{(2\pi\hbar)^3} M^{3/2} \sqrt{2} V \int_0^\infty \frac{\varepsilon^{1/2} d\varepsilon}{e^{\varepsilon/kT}-1}$$

$$= \frac{4\pi}{(2\pi\hbar)^3} (MkT)^{3/2} \sqrt{2} V \int_0^\infty \frac{x^{1/2} dx}{e^x - 1} \tag{6}$$

The integral has the value $1.306\ \pi^{1/2}$.

The whole treatment falls apart when $N(\varepsilon > \varepsilon_0) \sim N$: we define a critical temperature T_c such that at T_c $N(\varepsilon > \varepsilon_0) = N$: this temperature varies as $(\frac{N}{V})^{2/3}$. Below this temperature the number of atoms in the ground state varies as

$$N(\varepsilon_0) \approx N\{1 - (\frac{T}{T_c})^{3/2}\} \tag{7}$$

and below this temperature the specific heat varies as $\sim T^{3/2}$. T_c is the temperature below which the ground state occupancy increases enormously and almost all the atoms are to be found in this ground state: the Bose-Einstein condensation.

[PROBLEM: Estimate T_c for Helium at the density of liquid Helium (~ 0.12 gm cm^{-3}). Compare your answer with the temperature for the onset of superfluidity in liquid ^{4}He (2.2K).]

Above this temperature

$$N = \frac{4\pi}{(2\pi\hbar)^3} (MkT)^{3/2} \sqrt{2} V \int_0^\infty \frac{x^{1/2} dx}{e^{x-\mu/kT}-1} \tag{8}$$

which defines μ. The expression for the energy of the gas will be

$$E = \frac{4\pi}{(2\pi\hbar)^3} M^{3/2} (kT)^{5/2} \sqrt{2} V \int_0^\infty \frac{x^{3/2} dx}{e^{x-\mu/kT}-1} \tag{9}$$

and these integrals must be evaluated numerically in order to determine the behaviour between T_c and the classical region. [See for example Huang - Statistical Mechanics, Ch. 12, or Landau and Lifschitz - Statistical Physics.]

The Bose-Einstein condensation occurs in liquid ^{4}He which although a liquid has many of the properties of a gas (see Kittel, Ch. 17). The atoms are light and the interatomic forces weak. Everything else solidifies before the critical temperature is reached. However, superconductivity is associated with a Bose condensation of pairs of electrons which behave like bosons. [Both superconductivity and superfluidity are discussed in Feynman - Statistical Mechanics - but beware, this is very advanced work.]

[An interesting and illuminating PROBLEM is to show that there is no Bose condensation in a one or two dimensional gas. (See Kittel pp. 283, 284.) Before you do anything too formal examine the problem in terms of my treatment between eqs. (4) and (5).]

Thermal Radiation

We can treat the problem of thermal radiation very straightforwardly by considering a cavity at a temperature T full of a gas of photons which have energy $h\nu$ and momentum $h\nu/c$. The photons have a two-valued internal degree of freedom (electromagnetic radiation has two polarizations) but this corresponds to spin 1 with the state $m = 0$ (measured with respect to the direction of propagation) knocked out by the Lorentz condition (see Muirhead - The Special Theory of Relativity - Ch. 6).

The density of single photon states is thus given by

$$dn = 2 \times \frac{4\pi p^2 dp}{(2\pi\hbar)^3} V = 8\pi \frac{\nu^2 d\nu}{c^3} V \qquad (10)$$

which is of course the density of normal modes of the classical electromagnetic field in a resonant cavity. The individual photons do not interact directly (photon-photon scattering is negligible) but photons are continually absorbed and emitted by the walls of the cavity and it is this which maintains equilibrium at a temperature which is that of the walls.

The mean number of bosons in a single boson state of energy ε is

$$\bar{n} = \frac{1}{e^{(\varepsilon-\mu)/kT} - 1}$$

but this result was obtained assuming that the total number of bosons in the system of interest + reservoir is constant. This condition no longer holds when photons can be created and destroyed.

Remember how we set up the grand canonical treatment? We took as our definition of equilibrium the maximum term in the sum

$$\Sigma W(N, E)\, w(n,\varepsilon)$$

and found it by differentiating. The important term here is

$$w\frac{\partial W}{\partial N}\,\Delta N + W\frac{\partial w}{\partial n}\,\Delta n = 0 \quad \text{for a maximum.} \tag{11}$$

If the number of particles is conserved, then $\Delta N = -\Delta n$ and we obtained the condition

$$\frac{d\ell nW}{\partial N} = \frac{\partial \ell nw}{\partial n} \qquad (= -\mu/kT)$$

If, however Δn and ΔN are not linked, the only solution to (11) is given by $\mu = 0$.

Then for a gas of photons

$$\bar{n} = \frac{1}{e^{\varepsilon/kT}-1} = \frac{1}{e^{h\nu/kT}-1} \tag{12}$$

The number of photons between ν and $\nu + \Delta\nu$ is given by

$$n(\nu)\Delta\nu = \frac{8\pi}{c^3}V\frac{\nu^2\Delta\nu}{e^{h\nu/kT}-1} \tag{13}$$

and they carry energy

$$h\nu n(\nu)\Delta\nu = \frac{8\pi}{c^3}V\frac{h\nu^3\Delta\nu}{e^{h\nu/kT}-1} \tag{14}$$

[Now read Ch. 41 of The Feynman Lectures on Physics, Vol. I.]

For many purposes we are interested in the flux of radiation: the energy between ν and $\nu + \Delta\nu$ per unit area per unit solid angle. The propagation vectors corresponding to states in a cavity are spread uniformly over 4π steradians and the energy is travelling at velocity c. The energy density is obtained by dividing the energy by the volume in which it is contained, so

$$I(\nu)\Delta\nu\Delta\Omega = \frac{2}{c^2}\,\frac{h\nu^3\Delta\nu}{e^{h\nu/kT}-1}\Delta\Omega \tag{15}$$

(where the unit area is taken normal to the beam).

We may now obtain an expression for black body radiation - and can see at once that the energy radiated will be proportional to T^4! We are going to calculate Stefan's constant.

If you have a very tiny hole in the wall of a cavity and a little light gets in, it will never get out again: such a tiny hole is perfectly absorbing at all frequencies and so constitutes an ideal black body (or as near as we can get to one). Calculate the properties of the radiation emerging, by considering radiation propagating in a small element of solid angle at an angle θ to the normal to the surface of area ΔS

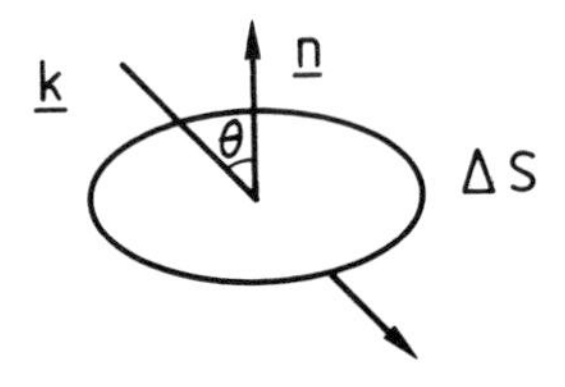

The area projected normal to the propagation vector $\underline{k}$ is $\Delta S\cos\theta$, an element of solid angle is $2\pi d\cos\theta$ so the energy radiated through the area Δs is given by

$$I(\nu)\Delta\nu\int_0^1 2\pi\cos\theta\, d\cos\theta\,\Delta S$$

$$= \pi I(\nu)\Delta\nu\Delta S$$

The energy radiated per unit area per second between ν and $\nu+\Delta\nu$ is thus

$$\frac{2\pi}{c^2}\frac{h\nu^3\Delta\nu}{e^{h\nu/kT}-1} \tag{16}$$

and the total energy radiated per unit area per second is

$$S = \frac{2\pi}{c^2}h\left(\frac{kT}{h}\right)^4\int_0^\infty \frac{x^3 dx}{e^x-1}$$

or $$S = \sigma T^4 \qquad \sigma = \frac{2\pi}{c^2}\frac{k^4}{h^3}\int_0^\infty \frac{x^3 dx}{e^x-1}$$

The integral has the value $\pi^4/15$ (see Mandl, Appendix A.2).
[PROBLEM: Calculate a value for σ and compare your answer with the standard value 5.6696×10^{-5} erg sec^{-1} cm^{-2}K^{-4}.]
[PROBLEM: Show that the radiation pressure of black body radiation is equal to $\frac{1}{3}$ of the energy density.]

In the early days of quantum theory the Planck distribution was used by Einstein to obtain an amusing and interesting result. If electromagnetic radiation is treated classically, but coupled to quantised atoms, the transition rates for absorbtion and induced emission are given by the same quantity, the square of a matrix element multiplied by the intensity of the radiation, say

$$BI(\nu)$$

taken at the appropriate frequency, $h\nu = E_2 - E_1$ where E_1 and E_2 are the energies of the relevant atomic levels. In equilibrium the number of excited atoms is $Ne^{-E_2/kT}$ and the number of ground state atoms is $Ne^{-E_1/kT}$. There are fewer atoms in the upper state than in the ground state, so for the transition rates to balance the atoms in the upper state must also have a spontaneous decay rate, such that

$$[BI(\nu) + A]Ne^{-E_2/kT} = BI(\nu)Ne^{-E_1/kT} \tag{17}$$

which yields

$$BI(\nu)[e^{h\nu/kT}-1] = A$$

$$I(\nu) = \frac{A/B}{[e^{h\nu/kT}-1]} \tag{18}$$

whence in order to obtain the Planck distribution

$$A \propto \nu^3 B \tag{19}$$

The spontaneous decay rate is given by ν^3 multiplied by the square of a transition matrix element calculated by treating the electromagnetic field as unquantised.
[See for example Schiff - Quantum Mechanics, Ch. 11 (3rd ed.).]

Do not forget that the universe is pervaded by black body radiation at a temperature of $\sim$3K, a relict of the big bang $\sim 10^{10}$ years ago.

Lecture 15

DEGENERATE FERMI GASES

The distribution function for the occupancy of single particle levels in a (non-interacting) gas of fermions is given by

$$\bar{n} = \frac{1}{e^{(\varepsilon-\mu)/kT}+1} \lesssim 1 \tag{1}$$

At absolute zero it is clear that if $\varepsilon > \mu$ $\bar{n} \to 0$ while if $\varepsilon < \mu$ the exponential $\to 0$ and $\bar{n} \to 1$. When most energy levels $\varepsilon < \mu$ have an occupancy of 1 and most energy levels $\varepsilon > \mu$ are unoccupied, we have a degenerate Fermi gas. The chemical potential μ is usually referred to as the Fermi energy, E_F.

There are a number of areas where the properties of degenerate Fermi gases are important. One is in the free electron theory of metals, where the conduction electrons can be treated for many purposes as a non-interacting Fermi gas confined with a box (the body of the metal). Another is liquid ^{3}He. Because the nucleus of ^{3}He has spin $\frac{1}{2}$, atoms of ^{3}He are fermions and because of the low mass and weak interatomic forces liquid ^{3}He has many of the properties of a degenerate Fermi gas. The electrons in white dwarf stars are degenerate, and so are the neutrons in neutron stars. Finally, the atomic nucleus has many of the properties of a gas of nucleons confined within a box - the nucleus itself.

As an example of a metal we will take copper. This is a metal with one valence electron and in the metallic state the valence electrons are more or less free to move throughout the metal. The first thing to do is to calculate the Fermi energy. The number of single particle states with momentum less than p is

$$2 \times \frac{4\pi}{3} \frac{p^3}{(2\pi\hbar)^3} V = \frac{8\pi}{3} \frac{(2m\varepsilon)^{3/2}}{(2\pi\hbar)^3} V \qquad (2)$$

where the factor of 2 is because of the two spin states of the electron. The Fermi energy is that energy below which all states are full at absolute zero, so

$$\frac{8\pi}{3} \frac{(2mE_F)^{3/2}}{(2\pi\hbar)^3} V = N \qquad (3)$$

The density of copper is 8.96 gm cm^{-3} and the atomic mass number is about 64. One gram of copper contains $\sim \frac{1}{64 \times 1.6 \times 10^{-24}} \sim \frac{6 \times 10^{23}}{64} \sim 10^{22}$ atoms so that one cm^3 contains $\sim 10^{23}$ free electrons. With $N/V \sim 10^{23}$

$$E_F = \left(\frac{3}{8\pi} \frac{N}{V}\right)^{2/3} \frac{(2\pi\hbar)^2}{2m} \simeq 9 \times 10^{-12} \text{ ergs} \simeq 6 \text{ eV} \qquad (4)$$

The quantity E_F/kT only becomes unity at 5×10^4°K (the Fermi temperature) so that any temperature where the metal is solid the gas of conduction electrons is highly degenerate, and the chemical potential is accurately given by the Fermi energy.

A small change in the temperature of such a gas can only change the state of a very small proportion of electrons near the top of the Fermi sea. We may define the thickness of the band where the occupation number is changing rapidly by some criterion like the range of energy over which $\bar{n}$ drops from $3/4$ to $1/4$ so that with

$$\frac{1}{e^{(\varepsilon-\mu)/kT}+1} = x \qquad (\varepsilon-\mu)/kT = \ln\left[\frac{1}{x} - 1\right]$$

and $\quad \Delta\varepsilon(3/4 - 1/4) = 2.2 \text{ kT}$

The number of electrons excited is thus $\sim \frac{NkT}{E_F}$ and each has an energy $\sim 2kT$ greater than at absolute zero and so the change in energy is

$$\Delta E \sim 2\frac{Nk^2T^2}{E_F}$$

and the specific heat is

$$C_V \sim 4Nk\left(\frac{kT}{E_F}\right) \tag{5}$$

The correct expression (rather than this rough estimate) can of course be calculated from the distribution function.

[PROBLEM: Do it! The lattice vibration contribution to C_V varies at low temperatures as T^3 (from Debye's model): make a comparison of the relative magnitudes of the electronic and lattice contributions as a function of temperature.]

[PROBLEM: Calculate the Fermi energy and Fermi temperature for nucleons in the nucleus ^{16}O, given that the nuclear radius is $r \simeq r_o A^{1/3}$ where A is the atomic mass number and $r_o \simeq 1.1 \times 10^{-13}$ cm. Significant applications to nuclear physics may be found in Fermi - Nuclear Physics, Appendix I.3, p22, Ch. VIII H, p159.]

For further applications of the theory of degenerate Fermi gases to metals you need a book on solid state physics. But we can use the above result for the heat capacity to make plausible without further calculation one remarkable result. The thermal conductivity will depend on the heat capacity of the electron gas multiplied by some characteristic transport velocity, while the electrical conductivity will depend on some characteristic transport velocity - we therefore expect the ratio of thermal to electrical conductivity to vary as T - the Wiedemann-Franz law, which was incomprehensible before the introduction of the theory of degenerate Fermi gases.

The pressure exerted by a degenerate Fermi gas is of great importance in the advanced stages of stellar evolution. In the core of a star the pressures are such that the electrons are not attached to any particular nucleus and form a gas of free electrons confined to the star. If the core has run out of nuclear fuel (either temporarily or permanently) the thermal energy will leak away and the thermal pressure will drop. However, even at absolute zero a Fermi gas exerts a pressure. The physical reason is simple. At absolute zero (or at any temperature $\lesssim T_F$) all electron states up to $\sim E_F$ are occupied. The energy of each state depends on the wavelength which in turn is governed by the dimensions of the enclosure. Reduce the volume of the enclosure:

this shortens the wavelength, increases the momentum and hence the energy of all the electrons, so you have to do work to reduce the volume of the gas.

We may calculate the pressure exerted by a degenerate Fermi gas by remembering that the pressure exerted by electrons with momenta between p and p + dp is given by (see Lecture 6)

$$dP = \frac{1}{3}n(p)pvdp \tag{6}$$

where $v = p/m$ and $n(p)dp$ is the number of electrons with momentum between p and p + dp per unit volume. For a degenerate gas the number of electrons between p and p + dp is given by the number of states between p and p + dp up to the Fermi energy, and then is zero. Thus if the electron energy ε is non-relativistic (as in metals)

$$\varepsilon = \frac{p^2}{2m} \quad \text{and} \quad v = \frac{p}{m}$$

so that

$$P = \int_0^{E_F} 2 \times \frac{4\pi p^2}{(2\pi\hbar)^3}dp\frac{1}{3}\frac{p^2}{m}$$

$$= \frac{8\pi}{3}\frac{1}{(2\pi\hbar)^3}\int_0^{E_F}(2m\varepsilon)^{3/2}d\varepsilon$$

$$= \frac{8\pi}{3}\frac{1}{(2\pi\hbar)^3}(2m)^{3/2}\frac{2}{5}E_F^{5/2}$$

while

$$\frac{N}{V} = \int_0^{E_F} 2\times\frac{4\pi p^2}{(2\pi\hbar)^3}dp$$

$$= 8\pi\frac{1}{(2\pi\hbar)^3}\int_0^{E_F}\frac{(2m)^{3/2}}{2}\varepsilon^{1/2}d\varepsilon$$

$$= \frac{8\pi}{3}\frac{1}{(2\pi\hbar)^3}(2m)^{3/2}E_F^{3/2}$$

so that

$$E_F = \left(\frac{3}{8\pi}\frac{N}{V}\right)^{2/3}\frac{(2\pi\hbar)^2}{2m}$$

and

$$P = \frac{2}{5}\frac{N}{V}E_F \qquad \text{(independent of temperature)}$$

or

$$P = \frac{2}{5}\left(\frac{3}{8\pi}\right)^{2/3}\left(\frac{N}{V}\right)^{5/3}\frac{(2\pi\hbar)^2}{2m}$$

It should be obvious that P is independent of temperature in a degenerate gas. In the core of a relatively low mass star, helium ignition may occur in conditions of electron degeneracy. The rate of helium burning rises very fast with temperature but provided that the thermal pressure of the nuclei is much less than the electron pressure the core won't expand until the temperature rises above T_F, by which time vast amounts of energy have been released - the core detonates rather than ignites. (The helium flash)

In hydrostatic equilibrium the internal pressure balances gravitational attraction at any radius and the equation of hydrostatic equilibrium is

$$\frac{dP}{dr} = \rho\frac{GM_r}{r^2}$$

where ρ is the mass density and M_r is the mass inside radius r. The central pressure and density are therefore related by

$$P_c \simeq \rho_c^2 \, GR^2$$

where R is the radius of the star. Setting this pressure equal to the degenerate electron pressure

$$\frac{2}{5}\left(\frac{3}{8\pi}\right)^{2/3}\left(\frac{N}{V}\right)^{5/3}\frac{(2\pi\hbar)^2}{2m} \approx \rho_c^2 GR^2$$

and we may express $\frac{N}{V}$ in terms of ρ_c. $\frac{N}{V}$ is the number density of electrons. For a helium core each electron is accompanied by two nucleons so $\frac{N}{V} = \frac{2MN}{V}\ \frac{1}{2M}$

$$= {}^{\rho_c}/2M \text{ where M is the nucleon mass}$$

$$\frac{1}{5}\left(\frac{3}{8\pi}\right)^{2/3} \rho_c^{5/3}\left(\frac{1}{2M}\right)^{5/3}\frac{(2\pi\hbar)^2}{m} \approx \rho_c^2 GR^2$$

The mass of the star is $M \approx \rho_c R^3$ so that

$$\alpha \left(\frac{M}{R^5}\right)^{5/3} \approx \frac{M^2}{R^4}\, G \qquad \text{where } \alpha = \frac{1}{5}\left(\frac{3}{8\pi}\right)^{2/3}\left(\frac{1}{2M}\right)^{5/3}\frac{(2\pi\hbar)^2}{m}$$

$$\approx 3 \times 10^{12}$$

or

$$R \approx \frac{\alpha M^{-1/3}}{G}$$

For a core of one solar mass $M_\odot = 2 \times 10^{33}$ gm,

$$R \approx 10^9 \text{ cm} \quad (R_\odot = 7 \times 10^{10} \text{ cm})$$

and $\rho \sim 10^6$ gm cm^{-3} ($\rho_{c_\odot} \sim 10^2$ gm cm^{-3}).
The Fermi energy is $E_F = 5M\alpha\rho^{2/3} \approx 0.1$ MeV and the Fermi temperature is $\sim 10^9$ K. Helium ignites at 10^8 K while hydrogen burns at a temperature $\sim 10^7$ K. (See for example Bowler - Nuclear Physics, Ch. 5.)

[PROBLEM: Consider a helium core of one solar mass, at a temperature $\sim 10^8$ K. Compare the degeneracy pressure, the thermal pressure of the helium nuclei and the radiation pressure. Are there any differences for a carbon core?]

A degenerate core which has shed its envelope of hydrogen will thereafter cool very slowly, supported against collapse by degeneracy pressure. Although hot, such a star is small and its luminosity is low because of the small surface area. Such stars are the <u>white dwarfs</u>.

But notice that $E_F \sim 0.1$ MeV, which is getting relativistic. For energies $\sim$ a few MeV, $\varepsilon \sim pc$ and $v \sim c$. Let us repeat the calculation for the relativistic regime.

$$P = \int_0^{E_F} 2 \times \frac{4\pi p^2}{(2\pi\hbar)^3} dp \, \frac{1}{3} pc \propto E_F^4$$

$$\frac{N}{V} = \int_0^{E_F} 2 \times \frac{4\pi p^2}{(2\pi\hbar)^3} dp \propto E_F^3$$

so $$\rho \propto E_F^3 \qquad P \propto E_F^4 \propto \rho^{4/3}$$

$$P \propto \rho^2 G R^2$$

and the equilibrium configuration yields

$$\rho^{4/3} \propto \rho^2 G R^2 \qquad \rho \sim \frac{M}{R^3}$$

So ρ and R cancel from the equation leaving a unique solution for M. Below this mass a degenerate star would expand until the bulk of the electrons were non-relativistic: above this mass, degeneracy pressure cannot support the star against gravitational collapse. The limiting mass is called the Chandrasekhar limit and is $\stackrel{\sim}{\sim} 1M_{\odot}$.

[PROBLEM: Put numbers into the above discussion and estimate the Chandrasekhar limit for helium (or carbon or iron...).]

[PROBLEM: Find the relation between mass and radius of a neutron star and find the Fermi energy and temperature for $M = 1M_{\odot}$. Can neutron degeneracy pressure sustain a more massive object than electron degeneracy pressure in a helium or carbon core can sustain? (Remember that for a neutron star approaching the relativistic domain the kinetic energy of the neutrons makes a significant contribution to the mass of the star.)]

Reference: Weinberg - Gravitation and Cosmology, Ch. 11.

Lecture 16

GRAVITATION

PART I - THE ATMOSPHERE

The gases we have considered so far have been gases confined by boxes. We now consider the simplest example of a gas confined by a gravitational field: an atmosphere. If the atmosphere has a scale height very much less than the radius of the planet, it should be a good approximation to represent the gravitational potential by gh where h is the height above ground. We suppose the atmosphere to be in equilibrium: consequently a region at the bottom of the atmosphere is (indirectly) in both diffusive and thermal contact with a region higher by h in the atmosphere and the temperatures and chemical potentials will be equal.

Imagine boxing these regions but connecting them by a tube to maintain thermal and diffusive contact. The energy levels in each box are determined by the walls as before, but these are levels of kinetic energy, ε_i. The total energy of any molecule, referred to a zero at the bottom of the atmosphere, is given by

$$E_i = \varepsilon_i + Mgh \tag{1}$$

The grand sum is given by

$$\mathcal{Z} = \Sigma\Sigma e^{[\mu N - E(N)]/kT}$$

$$\rightarrow \Sigma\Sigma e^{[N(\mu - Mgh) - \varepsilon(N)]/kT} \tag{2}$$

For a gas in a box,

$$\ell n \mathcal{Z} = \bar{N} = xe^{\mu/kT} \qquad \text{(Lecture 13)}$$

and since in (2) μ has been replaced by μ-Mgh it is clear that in any general region of the atmosphere we shall have

$$\ell n \mathcal{Z} = \bar{N} = xe^{(\mu - Mgh)/kT} \tag{3}$$

where $$x = \int e^{-\varepsilon/kT} \frac{dn}{d\varepsilon} d\varepsilon = yV$$

Thus the concentration $\bar{N}/V$ is given by

$$C = \frac{\bar{N}}{V} = ye^{(\mu - Mgh)/kT} \tag{4}$$

and since μ is equal at all points in the atmosphere, because it is in equilibrium,

$$C(h) = C(0)e^{-Mgh/kT} \tag{5}$$

where M is the molecular mass, and the pressure obeys the analogous relation

$$P(h) = P(0)e^{-Mgh/kT} \tag{6}$$

for one species of molecule.

The quantity $^{kT}/Mg$ is the scale height of the atmosphere for a species of mass M, which varies as $^{1}/M$.

[PROBLEM: Calculate the scale height of the Earth's atmosphere for molecular nitrogen and for atomic hydrogen.]

The results (5) and (6) can of course be obtained by elementary means, but the method we have used may be applied in other circumstances where a potential varies with position (for example, junctions in semiconductors). Note that the more massive a molecular species, the lower in an atmosphere it settles. This is of relevance to our second problem:

PART II - STAR CLUSTERS, GALAXIES AND CLUSTERS OF GALAXIES

These systems may be represented as a collection of classical mass points interacting through potentials

$$V_{ij} = -\frac{GM_iM_j}{r_{ij}} \tag{7}$$

Every mass point attracts every other mass point. If you consider a planet in circular orbit about a star, then the total energy is such that

$$\text{kinetic energy} = -\frac{1}{2}\ \text{potential energy} \qquad (8)$$

[PROBLEM: Prove it.]

This is a particular case of a more general result which applies to a system of gravitating mass points

$$\text{average total kinetic energy} = -\frac{1}{2}\ \text{average total potential energy} \qquad (9)$$

a result which is known as the virial theorem and is clearly part of statistical mechanics. It is easily proved. The potential energy is

$$V = \Sigma V_{ij} = -\frac{1}{2}\Sigma\Sigma_{ij} \frac{c_{ij}}{r_{ij}}, \quad c_{ij} = GM_iM_j \quad i \neq j$$

The equation of motion for the ith particle is

$$M_i\frac{d\underline{v}_i}{dt} = -\Sigma\frac{c_{ij}}{r_{ij}^3}(\underline{r}_i - \underline{r}_j) \qquad i \neq j$$

so

$$\Sigma_i M_i\underline{r}_i.\frac{d\underline{v}_i}{dt} = -\Sigma\Sigma\frac{c_{ij}\underline{r}_i}{r_{ij}^3}.(\underline{r}_i-\underline{r}_j) \qquad i \neq j \qquad (10)$$

Now

$$\frac{d}{dt}(\underline{r}.\underline{v}) = v^2 + \underline{r}.\frac{d\underline{v}}{dt}$$

and

$$\Sigma\Sigma\frac{c_{ij}\underline{r}_i.(\underline{r}_i - \underline{r}_j)}{r_{ij}^3} = -\Sigma\Sigma\frac{c_{ij}\underline{r}_j.(\underline{r}_i - \underline{r}_j)}{r_{ij}^3} \quad i \neq j$$

so (9) becomes

$$-\Sigma_i M_i v_i^2 + \Sigma_i\frac{d}{dt}(M_i\underline{r}_i.\underline{v}_i) = -\frac{1}{2}\Sigma\Sigma\frac{c_{ij}}{r_{ij}} = V$$

so that

$$\Sigma\frac{1}{2}M_i v_i^2 = -\frac{1}{2}V + \Sigma_i\frac{d}{dt}(M_i\underline{r}_i.\underline{v}_i) \qquad (11)$$

For a bound system AND ONLY FOR A BOUND SYSTEM the time average of $\underline{r}_i.\underline{v}_i$ will be zero, which proves (9) FOR A BOUND SYSTEM. We shall see that a set of gravitating mass points is not bound, but we can use the virial theorem for some limited time in considering a quasi-bound system. So

$$\bar{K} = -\frac{1}{2}\bar{V} \tag{9}$$

and
$$E = K + V$$

where K is kinetic energy.

$$\Delta E = \Delta K + \Delta V$$

so using (9)

$$\Delta\bar{E} = -\Delta\bar{K} \tag{12}$$

If we remove energy from a gravitating system, the mean kinetic energy increases - the system gets hotter. (A well known example of (12) is the increase in speed of an artificial satellite suffering atmospheric friction.) A physical illustration is the following: take some heat out of the sun. It will collapse, release gravitational energy and end up hotter - this is how the sun heated up to 10^7 K in the first place.

A cluster of gravitating mass points is not stable. Collisions among the members redistribute energy, energy can always be released by reducing the mean value of r, and the temperature then goes up as energy is lost. There is no thermal equilibrium, and such systems have NEGATIVE specific heats. In stars the collisions are atomic collisions and energy is lost in radiation. In clusters of stars the collisions are through the gravitational fields, and forming close sub-clusters releases gravitational energy and permits the evaporation of low mass members: in a thermal quasi-equilibrium, equipartition of energy, the heavier members sink and lighter members rise to the surface and may be evaporated.

The canonical approach may not be used for systems of classical self gravitating mass points because there is no equilibrium. If you imagine trying to establish an equilibrium between such a system and a reservoir, the system loses energy to the reservoir if the system is hotter, but this makes it hotter still. We have to work with some form of the microcanonical approach, and the statistical mechanics of the problem will be at best a quasi-equilibrium statistical mechanics.

Stars may be sustained in quasi-thermal equilibrium for a while by drawing on nuclear energy, and at a later stage of evolution this quasi-equilibrium may be converted into true equilibrium by degeneracy pressure (which is not a classical property) in white dwarfs or neutron stars.

But these forces cannot prevent gravitational collapse in stars of mass $\gtrsim 1.5\ M_\odot$ ...

Suppose a conglomerate of stars has N members and the average mass is m. The conglomerate occupies a volume V in space.

Let the average kinetic energy of a star be

$$\bar{k}_e = \frac{1}{2} m \bar{v}^2$$

The average potential energy due to a single second star is

$$\bar{U}_{ij} = -Gm^2 \left(\overline{\frac{1}{r_{ij}}}\right)$$

and

$$\left(\overline{\frac{1}{r_{ij}}}\right) \approx \left(\frac{N}{V}\right)^{1/3}$$

Then

$$\left|\frac{\bar{U}_{ij}}{\bar{k}_e}\right| \approx \frac{Gm}{\bar{v}^2}\left(\frac{N}{V}\right)^{1/3} \approx \frac{Gm}{\bar{v}^2}\frac{N^{1/3}}{R}$$

But from the virial theorem

$$\frac{1}{2} N m \bar{v}^2 = \frac{1}{2}\frac{GN^2m^2}{R}$$

So

$$\left|\frac{U_{ij}}{\bar{k}_e}\right| \approx N^{-2/3}$$

If N is a large number, then the kinetic energy of a star is large in comparison with the interaction energy with other individuals: motion is dominated by the mean field of the system. If N is small then of course the mean field is less important than local interactions, and close encounters dominate the motion.

The relaxation of a conglomerate to a quasi-equilibrium dominated by the mean field and satisfying the virial theorem is very rapid. Relaxation towards thermal equilibrium takes much longer. We can estimate the relevant relaxation time.

Let $\bar{v}$ be the mean relative velocity of the stars. The rate of collision is

$$\sim \bar{v}\sigma\frac{N}{V}$$

The collisions are long range gravitational collisions. The collision cross-section for impact parameter b is $\sim b^2$ and the momentum transferred in a collision is

$$\Delta p \sim \text{Force} \times \text{collision time} \sim \frac{Gm^2}{b^2}\frac{b}{v}$$

The energy transferred is $\sim(\Delta p)^2/2m$ so

$$\Delta k_e \sim \frac{1}{2m}\left(\frac{Gm^2}{bv}\right)^2$$

and the rate of loss of energy is

$$\text{collision rate} \times \text{energy lost} \sim (Gm^2)^2 \frac{1}{mv}\frac{N}{V}$$

The relaxation time $\tau \sim \dfrac{\text{energy}}{\text{rate of energy loss}}$ so

$$\tau \sim \frac{v^3}{G^2m^2}\frac{R^3}{N}$$

Put in some numbers:

	R(cm)	N	m(gm)	$\bar{v}$(cm sec^{-1})	$\bar{\tau}$(sec)	$\bar{\tau}$(years)
Globular Cluster	10^{19}	10^6	10^{33}	3×10^6	10^{19}	3×10^{11}
Galaxy	10^{22}	10^{11}	10^{33}	3×10^7	3×10^{25}	10^{18}
Virgo Cluster of Galaxies	10^{25}	2500	10^{44}	5×10^7	10^{20}	3×10^{12}

[PROBLEM: Check the values of $\bar{v}$ given against the virial theorem.]

The age of the universe is $\sim 10^{10}$ years. This is not much less than the mean thermal relaxation time of globular clusters, but galaxies as a whole exist in a quasi-equilibrium established not by innumerable close encounters but rather a quasi-equilibrium established in the mean gravitational field of the galaxy. (This too can be treated by statistical mechanics.)

We thus have reason to expect collapsed cores in globular clusters and perhaps in the dense cores of galaxies, while in large clusters the heavier galaxies are sinking into the middle. In the cores of globular clusters and galaxies, the heavier members sink and more tightly bound subclusters may form, releasing energy to the rest of the system (and cooling it) while getting hotter. Nothing can stop the eventual gravitational collapse of such a core, just as nothing can stop the gravitational collapse of a massive star. It is interesting that recent observations have yielded evidence for a very small very massive object of low luminosity, mass $\sim 5 \times 10^9 M_\odot$, radius <300 light years, in the core of the giant elliptical galaxy M87 which lies in the heart of the Virgo cluster of galaxies. M87 has a very active optical and radio-emitting jet extending 6000 light years - has the core of M87 collapsed to form a gigantic black hole?

[REFERENCES: Statistical Mechanics of Stellar Systems: Ch. 7 in Lectures on Selected Topics in Statistical Mechanics, D. ter Haar. (Pergamon, 1977) Rich Clusters of Galaxies: Scientific American 239, No. 5, 98 (November 1978).
On M87: Astrophysical Journal, 221, 721, 731 (1978).]

PART III - BLACK HOLES

A body so massive that degeneracy pressure will not support it contracts continually. The gravitational potential in the space outside it is

$$\phi = -\frac{GM}{r}$$

and if $r < R$ such that $\frac{GM}{Rc^2} \approx 1$ the binding energy of an object exceeds not only its kinetic energy but its total energy. For a photon escaping to infinity

$$h\nu_\infty = h\nu\left(1 - \frac{GM}{rc^2}\right)$$

and a photon is infinitely redshifted coming from the radius R, the Schwartzschild radius. The collapsing object disappears behind an event horizon and we have a black hole. The only information in the world outside that horizon is the mass of the hole, the angular momentum (if any) and the electric charge (if any). So lots of information is lost in the formation of a black hole: huge numbers of different configurations could lead to the same black hole. This suggests a black hole has a huge entropy. Could it be infinite - why not feed in an infinite number of very soft photons? You can't feed in a photon with $\lambda >> R$ so maybe the entropy is finite.

Lower an object on a rope to R, when its total energy is zero. Let it into the hole and haul back the rope. The rope going down turned a pulley and did work. The work done was

$$\frac{GmM}{R} = mc^2$$

the rest mass energy of the object.

So fill a box of side L with black body radiation at $r \to \infty$. Lower it to within a distance $^L/2$ of R (so the box doesn't disappear) and let the radiation out. Haul it back.

The net work done is

$$\frac{G\frac{\Sigma}{c^2}M}{R+L/2} \simeq (1 - \frac{L}{2R})\Sigma$$

The box must have $L >> \lambda$ (peak of black body spectrum at T) so the smallest value of L for which our box of radiation is macroscopic is $L >> {}^{hc}/kT$.

An amount of heat energy Σ is taken from a reservoir at temperature T and put to work between the reservoir and the black hole. The efficiency with which heat is converted into work with this heat engine is thus

$$\eta \lesssim 1 - \frac{hc}{2kRT}$$

and if we may apply thermodynamics to this situation

$$\eta \lesssim 1 - \frac{T_o}{T}$$

where T_o is the temperature of the sink. Then identify

$$T_o \sim \frac{hc}{2kR} = \frac{hc^3}{2kGM} \propto \frac{1}{M}$$

and attribute to the black hole this thermodynamic temperature.

The entropy gained by the black hole is

$$dS_o = (1 - \eta)\frac{\Sigma}{T_o} \gtrsim \frac{\Sigma}{T}$$

while the entropy lost to the reservoir is

$$dS = \frac{\Sigma}{T} \text{ and total entropy increases}$$

Writing $$dS_o = \frac{c^2 dM}{T_o} = \frac{2kGM}{hc} dM$$

leads us to identify the entropy of the black hole as a finite quantity

$$S_o \sim \frac{kG}{hc} M^2 \propto R^2$$

If a black hole has entropy and finite temperature, it should radiate like a black body ... but nothing can get out.

Hawking's famous work makes it very plausible that black holes would radiate like black bodies at the appropriate temperature T_0. Quantum field theory is needed for this. Outside R create a vacuum fluctuation: a positive energy and negative energy pair (measured with respect to $r \to \infty$). The positive energy one can escape, the negative energy one going down the spout and reducing the mass of the black hole, as seen from $r \to \infty$.

There is a connection between quantum theory, general relativity and one of the great pieces of physics of the last century - thermodynamics!

[PROBLEM: Calculate the temperature and entropy of a black hole of 1 $M_\odot$. Compare the entropy with that of the sun. How long will it take a 1 $M_\odot$ black hole to evaporate?]

REFERENCES FOR PART III

D.W. Sciama, Vistas in Astronomy, 19, 385 (1976).

P.C.W. Davies, Reports on Progress in Physics, 41, 1313 (1978).

J.D. Bekenstein, Physics Today, 33, 24 (1980).

FOR GENERAL INTEREST

F. Dyson, Time without end. Rev. Mod. Phys. 51, 447 (1979).

BIBLIOGRAPHY ON STATISTICAL MECHANICS

Some Basic Texts:	Referred to on pages

Baierlein, R. Atoms and Information Theory (486 pages, Freeman, 1971). — (47)

An introductory text, developing statistical mechanics from assumptions rooted in information theory. This is a new (only 20 years old) and interesting approach.

Kittel, C. Thermal Physics (418 pages, Wiley, 1969) — 27,48,58,77, 79,89,93

No compromises here! The canonical and grand canonical approaches are used right from the start, which is just fine except that these approaches seem so abstract when encountered first that it is difficult to keep track of the physics. But this is my favourite among the elementary texts. (Now issued in a revised edition by C. Kittel/H. Kroemer (Wiley 1980).)

Mandl, F. Statistical Physics (379 pages, Wiley, 1973) — 48,77,79,81, 90,96

A comprehensive undergraduate text, using primarily the canonical approach (see my remarks under Kittel).

Pointon, A.J. Introduction to Statistical Physics (202 pages, Longman, 1967)

A short and very traditional discussion.

Reidi, P.C. Thermal Physics (318 pages, Macmillan, 1976) — 81,84

A short and traditional discussion of statistical mechanics, but this book also treats both thermodynamics and kinetic theory.

	Referred to on pages
Rushbrooke, G.S. Introduction to Statistical Mechanics (333 pages, Oxford University Press, 1949) An ancient but useful book, good on applications but somewhat weak on fundamentals.	70,90

Advanced Texts:

Feynman, R.P. Statistical Mechanics (Benjamin, 1972) Chapter 1 alone provides a good introductory course, but at a very high level.	91,94
Fowler, R.H. Statistical Mechanics (Cambridge University Press, 2nd Ed., 1936). A classic work recently reprinted. Primarily concerned with applications; foundations developed using the Darwin-Fowler method of ensemble averages.	(27)
Huang, K. Statistical Mechanics (Wiley, 1963). An advanced and comprehensive text, but not too difficult.	84,93
Landau, L.D., Lifschitz, E.M., Pitaevskii, L.P. Statistical Physics (3rd Ed; 2 vols., Pergamon, 1980). An encyclopaedic work on equilibrium statistical mechanics.	84,93
Reichl, L.E. A modern Course in Statistical Physics (University of Texas Press; Arnold, 1980). Just that (ergodic theory, non-equilibrium processes, critical phenomena ...) Graduate level - frightening to the beginner.	
Tolman, R.C. The Principles of Statistical Mechanics (Oxford University Press, 1938) A classic work. Extremely interesting for an historical view of the foundations of the subject. For browsing.	

OTHER WORKS REFERRED TO

	Referred to on pages
Born, M. and Huang, K. Dynamical Theory of Crystal Lattices (Oxford University Press, 1954)	32
Bowler, M.G. Nuclear Physics (Pergamon, 1973)	103
Courant, R. and Hilbert, D. Methods of Mathematical Physics (2 vols., Interscience, 1953).	32
Fermi, E. Nuclear Physics (University of Chicago Press, 1950!)	100
Feynman, R.P., Leighton, R.B., Sands, M. The Feynman Lectures on Physics (3 vols., Addison-Wesley, 1964)	13,95
Kittel, C. Introduction to Solid State Physics (5th ed., Wiley, 1976).	37
Muirhead, H. The Special Theory of Relativity (Macmillan, 1973).	94
Schiff, L. Quantum Mechanics (3rd ed., McGraw-Hill, 1968).	97
ter Haar, D. Lectures on Selected Topics in Statistical Mechanics (Pergamon, 1977).	111
Weinberg, S. Gravitation and Cosmology (Wiley, 1972).	104
Whittaker, E.T. and Watson, G.N. A Course of Modern Analysis (4th ed., Cambridge University Press, 1963).	5

INDEX